THE POWER
OF NETWORKS

THE POWER
OF NETWORKS

Six Principles That Connect Our Lives

Christopher G. Brinton and
Mung Chiang

PRINCETON UNIVERSITY PRESS

PRINCETON AND OXFORD

Published by Princeton University Press,
41 William Street, Princeton, New Jersey 08540

In the United Kingdom: Princeton University Press,
6 Oxford Street, Woodstock, Oxfordshire OX20 1TR

press.princeton.edu

Jacket design by Karl Spurzem

Library of Congress Control Number 2016952294
ISBN 978-0-691-17071-8

British Library Cataloging-in-Publication Data is available

Typeset by T&T Productions Ltd, London

Printed on acid-free paper. ∞

Printed in the United States of America

10 9 8 7 6 5 4 3 2 1

To Kate and Sammy
  –Chris

To Novia, Augustan, and Vivia
  –Mung

Contents

Preface and Acknowledgments   ix

PART I   SHARING IS HARD   1

1   Controlling Your Volume   3

2   Accessing Networks "Randomly"   26

3   Pricing Data Smartly   44

A Conversation with Dennis Strigl   63

PART II   RANKING IS HARD   69

4   Bidding for Ad Spaces   71

5   Ordering Search Results   86

A Conversation with Eric Schmidt   101

PART III   CROWDS ARE WISE   109

6   Combining Product Ratings   111

7   Recommending Movies to Watch   128

8   Learning Socially   147

PART IV   CROWDS ARE NOT SO WISE   167

9   Viralizing Video Clips   169

10   Influencing People   188

PART V   DIVIDE AND CONQUER   209

11   Inventing the Internet   211

12   Routing Traffic   228

A Conversation with Robert Kahn   247

PART VI   END TO END   257

13   Controlling Congestion   259

14   Navigating a Small World   278

A Conversation with Vinton Cerf   299

Index   307

Preface and Acknowledgments

Networks are everywhere these days; even in places we least expect them. Everything from who we are friends with on Facebook to how our messages get carried through the Internet in a matter of milliseconds has interesting stories underlying it, whether a social network, a communication network, or an economic network. As different as recommending movies to watch, controlling device power levels, and viralizing video clips may appear, at the inner workings of these and other functions are six hidden principles that pop up in all different kinds of networks.

This book is about illustrating these principles, and the power of networks, in language that is accessible to anyone. Many great texts are already out there to expound the mathematical details and technical specifications of networks. This is not one of those.

Instead, this book describes the key ideas behind networking through storytelling, pictures, examples, and historical anecdotes including conversations with Google's Eric Schmidt, former Verizon Wireless CEO Dennis Strigl, and "fathers of the Internet" Vinton Cerf and Robert Kahn. Pictures supplement the text; there are hundreds of them within the pages. Analogies between networks and other parts of everyday life are drawn frequently. You might be surprised at just how similar the postal system, traffic jams, and stop signs are to Internet routing, network congestion, and WiFi random access, respectively.

This is not to say that the book doesn't show any math. Numerical examples can be very helpful in understanding the methods that drive networks. But the math you'll find in here is no more complicated than basic arithmetic: adding and multiplying numbers. All you need to read the book is a desire to learn.

How is this book organized? It's divided into six parts. Each part corresponds to one of the six principles of networking that connect our lives. Two or three chapters are included in each part on interesting topics that tell the story of that principle.

Now, what are these six principles? They are simple phrases that summarize a whole lot about the way networks are designed, built, and managed.

The first principle is: *sharing is hard* (part I). Whether you're making a call on your cell phone or browsing the Internet on WiFi, you have to share the network media, like the air, that you're using with many others. How is that possible without disrupting one another's connections? It requires effective techniques for sharing and coordination, from controlling the levels that our phones transmit at to pricing the data that we transmit.

Second, *ranking is hard* (part II). Many websites today have to resolve large banks of raw data to find an effective way to rank items. How do search engines like Google order the results we are shown? How do websites allocate ad spaces to advertisers? Ranking becomes even harder as the kind of items to rank becomes more complex.

The third principle is: *crowds are wise* (part III). Online retail and entertainment companies like Amazon and Netflix have lots of customers. Can we leverage the opinions of these large "crowds" to make product rating and recommendation more accurate and useful? The answer is yes, but we have to make certain assumptions about the crowd's opinions and how they were generated.

When these assumptions don't hold, it brings us to the fourth principle: *crowds are not so wise* (part IV). Why do video clips go viral? Because people can influence one another's actions and decisions. Do certain people have more influence in a social network than others? Yes, and in ways that are not necessarily intuitive, as you will see.

Divide and conquer is the fifth networking principle (part V). It is through this notion that the Internet is able to effectively scale up both its size functions—from routing to error correction—that it performs. Both the geographical and functional aspects of the Internet are divided intelligently into small pieces, so that each of them can be conquered separately.

The sixth and final principle, *end to end* (part VI), is about how a network operates over a large space. The devices we hold in our hands often don't know, nor do they need to know, what exactly is going on inside the Internet to perform the functions, like congestion control, that are assigned to them. Where in the network a job is carried out is an important question.

Throughout the text, you will see references to supplemental material that is on the book's website: www.powerofnetworks.org. Divided by chapter and organized into series of Q&A, this extra material is there in case you want to dig deeper on some of the examples, history, and more technical content beyond a basic understanding of the principles.

In addition to being a popular science book, what is contained between the covers and online can also be used as the basis for an introductory course in college or high school, offered to those interested in networking from any major or discipline. Instructional resources can be found on the book's website, or by emailing us at `learnPoN@gmail.com`. In fact, many of the materials for this book have already been used to teach more than 100,000 students in a Massive Open Online Course (MOOC) between 2013 and now.

We've had so much fun writing this book. We hope you'll enjoy reading it just as much!

Acknowledgments

I am indebted to several individuals who gave their feedback on different parts of this book: Bree, Hank, Kirsten, Loretta, Ray, Suzan, Vickie, and Yixin. That also goes for the many who proofread the initial version: Ethan, Harvest, Kate, Mo, Pranav, Rohan, Mom, Dad, and the diligent students in our MOOCs who detected typos. I also thank all the artists in my iTunes library for the music that kept me going through writing, especially Andy McKee, Eagles, Journey, Lynyrd Skynyrd, and Van Halen. Finally, and most importantly, this book would not have been possible without the unconditional love and support of my fiancée, family, and friends. A big THANK YOU to each of these outstanding individuals.

Chris, June 2016

Over the years, my teachers and students have taught me much about learning. I thank many colleagues who worked with me on topics related to networks, and the four visionary leaders who agreed to spend time interviewing with me. I am also grateful to the support from a Guggenheim Fellowship.

I keep borrowing time from my family to indulge myself on journeys of curiosity along diverse paths. My wife and my parents keep spoiling me by tolerating the deep overdraft. As to my children: Novia probably thinks this book, like all my speeches, is too long. Augustan would enjoy tearing the book apart. And Vivia, well, I'll be lazy and count this book as one of the gifts for her first birthday in November 2016.

Mung, June 2016

PART I
SHARING IS HARD

For a few seconds, sit back and imagine how your life would be without a cell phone. This may be difficult, since it would require you to revert to landlines every time you have to contact your boss or family, or have any conversation with anyone remotely. Then go back even further, to before you could have a telephone in your home. At this time, information could only travel as fast as you or a mailman could walk from street to street, as a train could travel from city to city, or as a freighter could cross the sea from continent to continent. Though hard to imagine, this is how people lived for thousands of years before wireline and wireless communications existed. These technologies made communication so much faster, while bringing about many interesting challenges to how network resources can be "shared" among so many people.

Here in part I of the book, we explore two types of wireless networks: cellular (in chapter 1) and WiFi (in chapter 2). When studying these networks, we will see two ways in which we share the resources (in this case, the electromagnetic spectrum over the air) by which we communicate. Managing interference is really the key: it requires methods to figure out when you communicate and how loudly you communicate, completely behind the scenes.

Network pricing (chapter 3) can also be an effective way of enabling more efficient resource sharing. We will see the methods network providers use to determine how much we pay for what we consume.

1

Controlling Your Volume

Cell phones are now part of our daily lives. Take illustration 1.1, which shows the **mobile penetration** for selected countries in mid-2015. This is the average number of cell phones per person in the country. Notice that the leftmost five are over 100%, meaning there are more cell phones than people in these cases.

Also, the 13 countries in illustration 1.1 were the ones in mid-2015 with over 100 million subscriptions. If we took the worldwide total number of purchased cell phone subscriptions at this time, it was over 6,800,000,000 (that's 6.8 billion!).

With these enormous numbers, you might ask, how is it possible that we can all communicate in the air effectively without disrupting one another's calls, messages, or Internet usage? We'll see a couple of methods for sharing in this part of the book, starting with power (speaking volume) control in this chapter.

The modern mobile **cellular** system is the result of decades of technological innovation. As mobile devices moved from being a luxury item in the 1940s–80s to an utter necessity by the twenty-first century, engineers had to come up with different ways of letting people share the air.

FROM TELEPHONE TO CELL PHONE

Before the advent of wireless networks and cell phones, communication networking was more about **wireline**. Wireline is communication using wires, as opposed to **wireless**. The first phone call was made all the way back on October 9, 1876, by Alexander Graham Bell, over a 2-mile wired stretch from Boston to Cambridge. The following year, Bell founded the Bell Telephone Company, the first company that provided a **public switched telephone network** service (we usually refer to this as "landline").

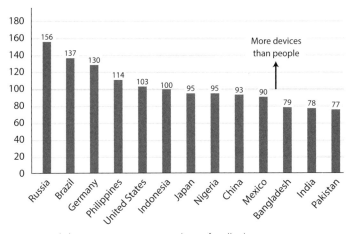

Illustration 1.1 Mobile penetration, or number of cell phones per person, in selected countries as of June 2015. Six countries have a penetration of at least 100%, meaning there are more phones than people.

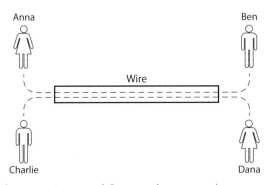

Illustration 1.2 Suppose Anna and Ben are having a phone conversation, as are Charlie and Dana. How is it possible for both pairs to use the same wire without interfering with each other?

Before Bell designed the telephone, he was experimenting with the telegraph, an earlier invention. The "multiple telegraph" was one where multiple **transmitters** (senders of messages) and **receivers** (receivers of messages) could operate over a single wire.

Stop there for a second: how is it possible for us to have many different people sharing the same wire, as in illustration 1.2? If Anna and Ben are trying to talk to each other, as are Charlie and Dana, wouldn't this cause them to disrupt each other's conversations?

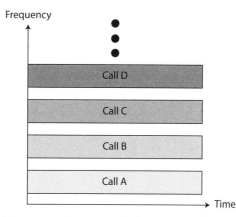

Illustration 1.3 With frequency division multiple access, calls are distinguished by their designated frequency channels: the users on Call A get one channel, those on B get a different one, and so on.

Not necessarily. Even though they are sharing the same *space* (the wire), we can separate them along some other dimension. The most intuitive one is perhaps time: let Anna and Ben use the wire for a bit, then Charlie and Dana, then back to Anna and Ben, and so on. We could also try language: have Anna and Ben talk in English, and Charlie and Dana talk in Spanish. Then they can just listen for their own language while still talking at the same time. In this case, we still have to worry about different voices overpowering each other.

These dimensions—time and language—are simplified examples of different **multiple access** technologies. These techniques allow multiple users to share the same network medium (e.g., wire or air). We take a look at them in more detail throughout this chapter.

Sharing by Frequency

The multiple telegraph separated conversations along different *frequencies*, in what is known as **frequency division multiple access**, or FDMA. FDMA allocates each transmitter-receiver pair, which we call a link, a separate **frequency channel** by which they can communicate. You can see this in illustration 1.3.

What is a "frequency"? For the ones we can hear, we distinguish them as different pitches of sound. Frequency is measured in units of **hertz** (Hz), which indicates the number of cycles per second in a wave. So 10 Hz means

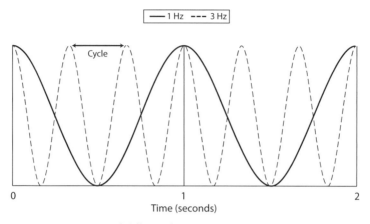

Illustration 1.4 Two waves of different frequencies, which is the number of cycles the wave completes in one second. The solid wave has a frequency of 1 Hz, while the dashed wave has a frequency of 3 Hz.

that a wave completes 10 cycles per second (see illustration 1.4). For more on frequency channels, check out Q1.1 on the book's website.

The unit of frequency will be mentioned many times in this part of the book, but the actual ranges we will deal with are many orders of magnitude higher than a hertz. Wireless frequency bands are typically referenced in millions of hertz, known as megahertz (MHz), or billions of hertz, known as gigahertz (GHz). To give you an idea of how large these are, the highest frequency a human can hear is about 20,000 Hz.

The first mobile phones, dating all the way back to the 1920s–30s, used FDMA. They were **analog** in nature, meaning that their signals traversed the air in their exact electrical forms. In 1946, the first "mobile-phone" network, called the Mobile Telephone Service, was introduced by Bell Telephone. This was an FDMA system, as was its successor in 1964. They are considered zeroth generation, or **0G**, technologies, as opposed to the **4G** technologies that we use today.

The First Handheld

Back in the 1970s, Martin Cooper of Motorola was convinced that handhelds were the wave of the future. In 1973, he and a team spent 90 days making the first-ever mobile handheld phone: the DynaTAC.

The DynaTAC was not what we view today as a handheld. Weighing close to 2 lbs, it cost almost $3,000 (in 1973 dollars!) and offered 30 minutes of call time before charging was necessary. By comparison, an iPhone in 2016 weighed less than one-third of a pound, cost as low as $150 (depending on the model and the wireless contract), and offered many hours of voice and data applications after each recharge.

It wouldn't be until the mid-1990s that the industry would really start switching from car phones. Similar to digital networks, palm-sized phones only became practical once the cost of electronic components began to reduce drastically, which is in turn partially driven by the volume of demand, which is in turn partially driven by the applications enabled by such technologies.

The "Cell" in "Cell Phone"

In 1976, New York City alone had about 500 mobile subscribers, and more than six times this number on the waiting list. There was a dire need to increase network **capacity**. So what could network operators do? There were really only two options: petition the Federal Communications Commission, or FCC, for more spectrum, or figure out a way to fit more users into the same spectrum.

For more information on the FCC licensing process, check out Q1.2 on the book's website. How can we put more users in the same spectrum? Maybe we can reuse the channels? That seems kind of far-fetched: if we have two links right next to each other, and they are using the same channel, certainly they would interfere. But what if they aren't right next to each other? If they are far enough away, then could we reuse the same channel?

The answer is yes. When signals propagate through the air (as well as through wires), their power levels **attenuate**. This means that they diminish with further distance, as in illustration 1.5.

Typically, attenuation is looked on as a negative quality. It causes a signal to weaken, making it harder to transmit over long distances. But that's exactly what we need here: if you and I are far enough apart, we can each make a call without overlapping in space.

The nature of attenuation led engineers to begin dividing mobile regions geographically into **cells**, often represented as "hexagons" (for good reason, beyond our scope here). The idea is that any given cell can be assigned a set of frequencies not being used by a cell adjacent to it. In

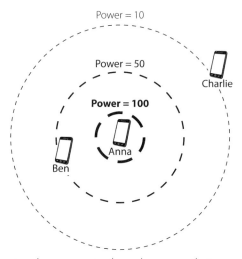

Illustration 1.5 As signals propagate through space, their power levels attenuate. Around Anna's phone, her transmit power level is 100. By the time the signal gets to Ben, it's 50, and at Charlie, it's 10.

this way, cells that are using the same channel will be far enough away from each other that it won't matter, allowing us to be more efficient with the available resources.

You can see an example of a cell diagram in illustration 1.6. Here, any of the hexagons with the same shade will be assigned the same frequencies, since they are not adjacent. Let's say the darkest shade gets channels 1–4, the middle shade gets 5–8, and the lightest shade gets 9–12. Rob is in a dark cell, and is on channel 2. Someone else in his cell may have 1, 3, or 4. Since Rachael is in a different dark cell, she can also be assigned channel 2, because she is far enough away. Ben, in a middle cell, cannot get channel 2, because he is too close to the dark cells. When we assign colors (frequencies) to cells, we often want to use the smallest number of colors possible. Finding that combination can actually be quite difficult, especially as the number of cells in the diagram gets very large.

So, what's in a cell? There are **base stations**, or BSs, and **mobile stations**, or MSs. The BSs in each cell are connected on one side to the wired core network and the Internet, and on the other to each mobile station assigned to it. An MS could be a cell phone, tablet, or any device that can transmit and receive according to a cellular standard.

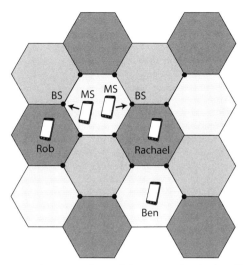

Illustration 1.6 This is a diagram of a cellular network. Each cell is a hexagon, with multiple mobile stations (MSs) and base stations (BSs). The shading of a cell indicates the frequency band that the cell is using. No two neighboring cells have the same shading, and so use different frequency bands to prevent interference.

Cells were first used in the advanced mobile phone system, which marked the beginning of **1G** technology in the United States. Under this system, the number of mobile subscribers skyrocketed. By the 1990s, there were 25 million cellular subscribers in the United States alone. This also meant that, due to the high usage rate and low capacity, analog just could not cut it anymore.

ENTER DIGITAL

With analog networks once again overcrowded, the United States and other countries began experimenting with an alternative: **digital** systems. An analog signal will be "digitized" by converting it into a sequence of **bits**, or 1s and 0s (see illustration 1.7).

Digital systems can offer enormous advantages in terms of capacity, as they enable use of two other multiple-access technologies that we discuss next. Prior to the late 1980s, the small-scale electronics necessary to realize these networks were just not yet available at a low enough cost.

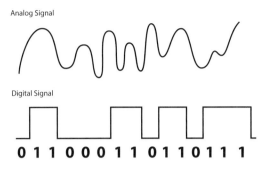

Illustration 1.7 An analog signal is one that changes continuously in time. In contrast, a digital signal is a series of bits, or 1s and 0s.

Sharing by Time (and Frequency)

The transfer from analog to digital cellular marked the migration from 1G to **2G**. The first 2G standard was the global system for mobile communications, or GSM, which began in 1982. By 1987, it was able to achieve three times the capacity of analog.

Digital coding allows us to compress multiple conversations into one band. So, even within one cell, we can have a number of people sharing the same frequency channel. We just have to add another one of our dimensions into the mix. The most obvious choice for the extra dimension is time.

In other words, multiple users can share the same frequency channel, but they all have to take turns. Each is allocated a different timeslot, in a scheme known as **time division multiple access**, or TDMA. You can see an example of TDMA in illustration 1.8.

GSM was adopted in much of Europe rather quickly, as the European Union favored the development of one common standard. GSM is still used in parts of the world today, operating mainly in the 900 and 1,800 MHz frequency bands. It drove down the cost of phones and marked the transition to handhelds that offer texting, gaming, and other entertainment.

Sharing by Codes

The story of 2G standard adoption in the United States is even more interesting. With the knowledge of increasing capacity demands, the US Cellular Telecommunications Industry Association posted a set of performance

Frequency

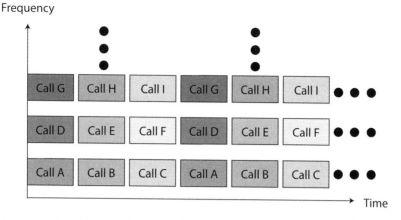

Time

Illustration 1.8 With time division multiple access, a certain number of calls (in this case three) can share the same frequency channel. For instance, calls A, B, and C are assigned the same channel but are separated in time.

Frequency

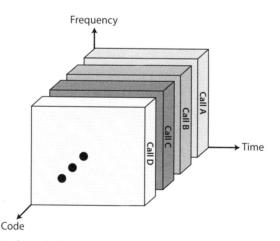

Time

Code

Illustration 1.9 With code division multiple access, calls are distinguished along the ``code'' dimension. All calls may operate over the same frequencies and at the same times, because each transmission in the network is assigned a unique code.

requirements in 1988 that the industry should aim to meet with the first digital cellular standard. The main one was a ten-fold improvement in capacity over legacy analog networks.

By this time, virtually every network operator and device manufacturer in the United States felt TMDA was the best way to go. But not the company

Qualcomm. They championed another technology, **code division multiple access**, or CDMA. With CDMA, users are separated over the "code" dimension and are undistinguished in both time and frequency, as shown in illustration 1.9. The best analogy to a code is perhaps language: it's as if you gave each link a different one to speak.

Each code is like a key. The sender locks the message, sends it out, and only gives the key to the receiver. The difficulty in designing such codes is that only one key should be able to "unlock" any given signal. If another receiver tries to descramble this message using its own key, it should appear as noise. A collection of codes possessing this property, where each one "cancels" the other, is referred to as a family of **orthogonal codes**. For more information on how CDMA works, check out Q1.3 on the book's website.

An early prediction claimed that CDMA could provide a capacity improvement of 40 times over that of legacy analog networks. Despite this claim, most of the engineers, manufacturers, and operators at the time resisted CDMA. For one, there had not yet been a demonstration of CDMA in a prototype cellular network.

In 1989, the Cellular Telecommunications Industry Association voted and approved TDMA as the first 2G digital standard in the United States. It would take more proofs-of-concept over the next 4 years before CDMA would be approved.

THE COCKTAIL PARTY ANALOGY

Here's a useful analogy that illustrates some of the technologies that we have introduced up to now. Imagine a cocktail party taking place in a large mansion with many rooms, in which there are many conversations occurring. Given that there are a lot of people at this party, if everyone was crammed into the same room and talking simultaneously, it would be difficult to hear what was going on in your own conversation. We leave it up to the host to determine the best way to manage this.

The host first decides that there can be one couple having a conversation in each room. Each couple has their room until their conversation is finished, so each person can speak at a comfortable volume, because voices will have attenuated by the time they reach the other rooms. But if we think of rooms as cells, this would be like only allowing one link per cell at a time.

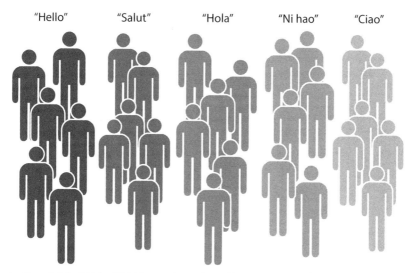

Illustration 1.10 With CDMA, each code is like a separate language. In the cocktail party analogy, multiple conversations can occur in a room if they use different languages. The issue then becomes controlling speaking volume levels.

Given that there are probably many more guests than rooms, this would not be desirable to many of the couples who were not assigned one.

To deal with this capacity issue, the host decides to allow many couples to share each room (i.e., many people per cell), telling each pair to speak at separate times. So, in any given room, the first pair has maybe 30 seconds, while everyone else remains silent, followed by the next group, and so on. Again, each person can speak as loud as she wants, because conversations will not overpower one another. This is an example of TDMA, where in each room every conversation is assigned a separate timeslot.

Rather than assigning timeslots, suppose the host asks each pair in a room to use a separate language. Then, everyone can speak simultaneously, since each pair is listening for one language in particular. This is an example of a CDMA system, where each language represents a different code (see illustration 1.10). But human languages were not designed as perfect codes. Further, volume control becomes an issue here, because everyone in a room can hear all the other conversations, regardless of what language is spoken. We would require some type of coordination, where individuals would adjust their volumes based on their distances from one another.

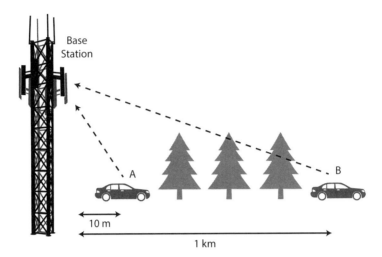

Illustration 1.11 The farther a transmitter is from its receiver, the higher the attenuation is, and the more objects there are to obstruct the path. Here, A has a short, clear path to the tower, while B has a long path that is obstructed by objects (e.g., trees).

CONTROLLING POWER LEVELS

CDMA had its share of complications. We now explore some of the major problems its advocates had to overcome in the early 1990s.

Near-Far Problem

Whenever signals are traveling at the same time, there is inevitable **interference**. The problem becomes worse when you take distance from the BS into consideration. How can someone who is a mile from the BS make a call without it being ruined by someone only a few meters from it? Not only does this person face more attenuation, but also there are likely more objects (like trees) to obstruct the signal's path. This leads to differing levels of **channel quality**, as in illustration 1.11.

What we're describing here is known as the **near-far problem**. To deal with it, our phones need some mechanism by which we can adjust our transmission powers to compensate for differences in channel quality, so that we can share the air effectively.

The initially proposed solution to mitigate this was the **transmission power control** (TPC) algorithm. It was an attempt to equalize received signal powers. The BS would measure what it was receiving from each

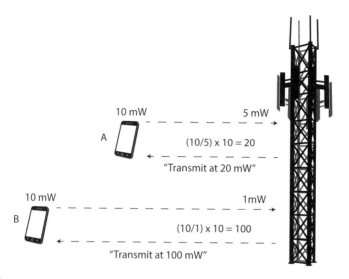

Illustration 1.12 Example of the transmission power control algorithm.

transmitter, compare this with the desired power, and send a feedback message to each device, telling it to adjust accordingly.

How is power measured at a receiver? The standard unit is the **watt** (W), which is the amount of energy transmitted per second. So, 5 W means that five units of energy are transferred per second. In this chapter, we will typically be dealing with power levels that are many fractions of a watt, either milliwatts (mW), which are thousandths of watts, or microwatts (μW), which are millionths.

Back to the TPC algorithm. Let's say the desired power level at the tower is 10 mW. Two cell phones, A and B, start sending at this power, and the BS receives them at 5 mW and 1 mW, respectively, as you can see in illustration 1.12. Channel degradation caused A's power to be halved and B's power to be reduced by a factor of 10. To "reverse" this, TPC asks the transmitters to send at twice and ten times their current transmission powers, respectively. That means A should send at 2 × 10 mW = 20 mW, and B should send at 10 × 10 mW = 100 mW.

More generally, the TPC algorithm is based on this equation:

$$\text{Next power} = \text{``The ratio''} \times \text{Current power}$$

where "The ratio" is the desired power (10 mW in this example) divided by the received power (5 and 1 mW).

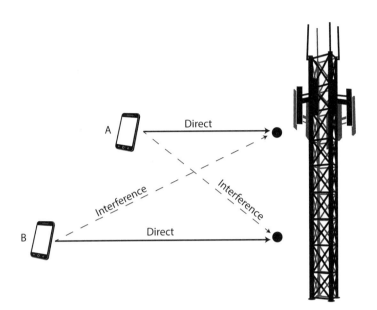

Illustration 1.13 Ideally, only the power from the transmitter of a link would be present at its receiver. But this is not the reality: here, some of A's transmission will be coupled into B's receiver, and vice versa.

Quality Is More Than Power

With TPC, the objective is to equalize the received signal power by boosting the transmissions accordingly. Is this enough to ensure "good reception," though? Not necessarily. The received signal is also going to be impacted by interference from other phones. You can see this in illustration 1.13. Even if the transmission power on link A is high, if the interference coming from other transmitters (link B) is also high, then A's received signals may still have low quality. This is our first glimpse of the impact of the network; in this case, the impact is on multiple users accessing the same communication medium.

For mobile communication, it is typically quality, rather than power, that we need to equalize. So, how do we determine the received signal quality on a link? We can view it as a combination of three factors:

1. The received signal power from the desired transmitter. This is the transmitter the receiver is trying to listen to.

2. The received signal power from the undesired (i.e., interfering) transmitters. These are the transmitters the receiver hopes to avoid.

3. The receiver noise, which is something inherent in every receiver.

Quality is measured as the good (item 1) divided by the bad (items 2 and 3). It's called the **signal-to-interference ratio** (SIR).

Preventing an Arms Race

Achieving target SIRs is more complicated than achieving target powers, because we cannot simply increase transmission powers to achieve the desired SIRs simultaneously. Increasing A's transmission power will increase A's SIR, but it will also cause the SIRs of other links to decrease. We would then have to increase the other transmitter powers to raise their SIRs, which would affect everyone else, causing them to increase, and so on. The result would be an inevitable "arms race," ending in whoever could transmit at the highest power winning. That wouldn't be a very effective way to share.

If each mobile device fixes a desired SIR, is it possible to find a set of transmission powers that will meet all of the targets simultaneously? The answer is yes, as long as the desired SIRs are **feasible**, or mutually compatible with one another. In other words, there cannot be a collective desire for unrealistically high SIRs.

The solution is known as **distributed power control**, or DPC. It works like this:

1. The devices each start with some initial transmission power.
2. The receiver measures the SIR for each transmitter.
3. Based upon the ratio between the target and the measured SIRs, each transmitter adjusts its power level.
4. Repeat steps 2 and 3 as needed.

DPC is an **iterative** algorithm that repeats over and over, unlike the single step, near-far TPC algorithm discussed earlier. Given that the target SIRs are feasible, it turns out that the DPC algorithm will **converge**, meaning that the SIRs will be met by power levels that will stop updating. In fact, these convergent power levels will also be **optimal** ones, in the sense that they will use the least amount of energy.

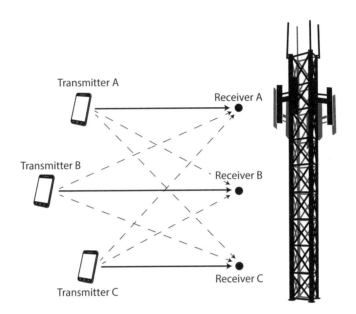

Illustration 1.14 Channels between three mobile stations and the base station. The solid lines denote direct channel gains, while the dashed lines represent the interference channel gains.

The Workings of Distributed Power Control

Think of three MSs, A, B, and C, in a single cell, as in illustration 1.14. Here, the solid lines represent the direct **channel gains** uplink for each of the transmitter-receiver pairs. Channel gain is a measure of how much the power is **amplified**, or, for all practical purposes, multiplied, from source to destination (as these are fractional quantities, the multiplication is really attenuation). Direct channel gain should be as high as possible, because it represents an intended conversation. In contrast, the dashed lines represent the interference channel gains, or by what factor the unintended signals will couple to each receiver. You can see the channel gains, the target SIRs, and receiver noise for our example in illustrations 1.15 and 1.16. (These numbers are meant for illustrative purposes, and don't represent actual numbers typically observed in cellular networks.)

Let's illustrate the DPC algorithm's computation for its first few steps. To update the transmission powers, the DPC algorithm employs an intuitive equation similar to the near-far TPC equation four pages earlier, but

Transmitter	Receiver of link		
of link	A	B	C
A	0.9	0.1	0.2
B	0.1	0.8	0.2
C	0.2	0.1	0.9

Illustration 1.15 Channel gains for our distributed power control example.

Link	Target SIR	Noise (mW)
A	1.8	0.1
B	2.0	0.2
C	2.2	0.3

Illustration 1.16 Desired signal-to-interference ratio (SIR) and noise parameters in our distributed power control example.

involving SIRs rather than channel quality. For each transmitter, the update is

$$\text{Next power} = \text{``The ratio''} \times \text{Current power}$$

where "The ratio" is now the desired divided by the measured SIR.

This update is quite logical. If the measured SIR is lower than the desired SIR, then "The ratio" will be larger than 1. The transmission power will then increase, in an attempt to equalize them. On the contrary, if the measured SIR is higher than the desired one, "The ratio" will be lower than 1, and the transmission power will decrease. This transmitter can afford to use less power, and this action will help improve the SIRs for the other transmitters, too. Finally, if the measured and desired SIRs are the same, "The ratio" will be 1, and the transmission power will stay the same. There's no need to change if the target is already being met.

Why is such an update necessary? In a cell, each device imposes **negative externality** on the others, by interfering with them. In other words, while achieving its own benefit, a device does some "damage" to the rest of the network as well. This update keeps the devices in check: whenever the SIR is higher than it needs to be, the power level is dropped, and whenever it's too low, the power level is raised.

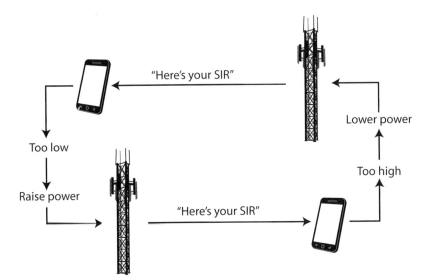

Illustration 1.17 The tower tells each device its current received signal-to-interference ratio (SIR), which serves as a negative feedback signal. Using this, each device can update its transmission power independently.

The process of "message passing" between the BS and the devices to correct for such deviations is an example of **negative feedback** (see illustration 1.17). It forces the transmitters to internalize their negative externalities (i.e., to pay for the interference they cause) by following the rules to make up for their added interference to the system.

The concepts of negative feedback and negative externality will come up repeatedly as we look at different networks in this book. More generally, negative feedback is a way of maintaining **equilibrium** in a system by checking for and counterbalancing fluctuations in the output. Positive feedback, where we move away from equilibria, will come up later, too.

Back to our example. Of all the quantities needed in the DPC equation, we know the desired SIR values (from illustration 1.16), and we can start off the current transmission powers at 2 mW. That leaves the measured SIRs. For each link, we can calculate this as

$$\text{Measured SIR} = \frac{\text{Signal}}{\text{Interference} + \text{Noise}}$$

How do we get "Signal," "Interference," and "Noise"? Let's start with Link A:

- *Signal*: This is the direct gain from Transmitter A to Receiver A, multiplied by the transmission power. From illustration 1.15, it is 0.9×2 mW = 1.8 mW.
- *Interference*: This is the sum of the indirect gains from other transmitters to Receiver A, multiplied by their transmission powers. In illustration 1.15, we look at the first column: from B to A, the gain is 0.1, and from C to A it is 0.2. That gives 0.1×2 mW + 0.2×2 mW = 0.6 mW.
- *Noise*: This is the receiver noise, given in illustration 1.16 to be 0.1 mW for Link A.

In reality, the receiver does not even need to do this multiplication and addition, because it can physically measure the SIR.

So, the measured SIR for Link A is

$$\frac{1.8}{0.6 + 0.1} = \frac{1.8}{0.7} = 2.57$$

What about Link B? From Transmitter B to Receiver B, the direct gain is 0.8, so the signal power is 0.8×2 mW = 1.6 mW. The indirect gains come from Transmitters A and C: from A to B, the gain is 0.1, and from C to B it is also 0.1. That means the interfering power is 0.1×2 mW + 0.1×2 mW = 0.4 mW. Finally, the receiver noise for Link B is 0.2 mW. So, the measured SIR for Link B is 1.6/0.6 = 2.67.

Using the same procedure, you can find the SIR for Link C to be

$$\frac{0.9 \times 2}{0.2 \times 2 + 0.2 \times 2 + 0.3} = \frac{1.8}{1.1} = 1.64$$

Let's compare these values to the desired SIRs. For Link A, the desired value is 1.8, and so the measured SIR is too high by $2.57 - 1.8 = 0.77$. Similarly, the measured SIR for Link B is too high by $2.67 - 2.0 = 0.67$, while that of Link C is too low by $2.2 - 1.64 = 0.56$.

We can now calculate the new power levels using the DPC equation. What are the ratios? We divide the target by the measured SIRs for each link to get these: 1.8/2.57 = 0.70, 2.0/2.67 = 0.75, and 2.2/1.64 = 1.34. As expected, the ratio is less than 1 for Links A and B (the measured value is too high), and greater than 1 for Link C (measured value is too low). With these values, the next power levels for A, B, and C are

$$0.70 \times 2 \text{ mW} = 1.40 \text{ mW}$$

$$0.75 \times 2 \text{ mW} = 1.50 \text{ mW}$$

$$1.34 \times 2 \text{ mW} = 2.68 \text{ mW}$$

respectively. Negative feedback has caused the power levels of A and B to decrease, and that of C to increase, as we expected.

What is the next step? To calculate the SIRs at these new power levels, we use the equation on page 20. And after that? We adjust the power levels based on the update equation on page 19. The calculations for future steps are carried out in the same manner. If you'd like to see more done by hand, check out Q1.4 on the book's website.

The transmission powers and SIR levels for 30 iterations of the DPC algorithm are graphed in illustration 1.18. After roughly 10 iterations, we can no longer see any noticeable changes in either quantity, indicating that the DPC algorithm has converged to an equilibrium. The measured SIRs have reached their target values of 1.8, 2.0, and 2.2, with power levels of 1.26, 1.31, and 1.99 mW, respectively.

Why does Link C have a much higher power level than the other two? It has the highest noise component of any receiver (0.3 mW), the highest interference gains from other links (both 0.2), and the highest target SIR value of any link (2.2). It needs a higher transmission power to overcome these disadvantages.

Why does the algorithm converge? Since the measured SIRs are the same as the target SIRs, "The ratio" in the update equation will be unity, so the power will not change any more. Negative feedback has brought the network to an equilibrium at which the devices are sharing effectively. It will stay this way until the network changes, like when a device's interference conditions change, a new device enters the cell, or an existing device leaves the cell.

In a real cell, there can be hundreds of phones, and as conversations are started and stopped and people move from location to location, you can imagine that the channel conditions and SIR values from link to link will change quite rapidly. As a result, it is required to have power control implemented up to 1,500 times each second. One benefit of the DPC algorithm is that each device really doesn't need any knowledge of how the other links are operating. To calculate its next power level, all it needs is its current transmission power, target SIR, and current measured SIR. These are all its own parameters, and it makes its current decision independently (e.g., there's no need for it to know the SIRs of any other links). This allows each device to perform its computations internally, without the need to share information with the others. In other words, the DPC is a completely

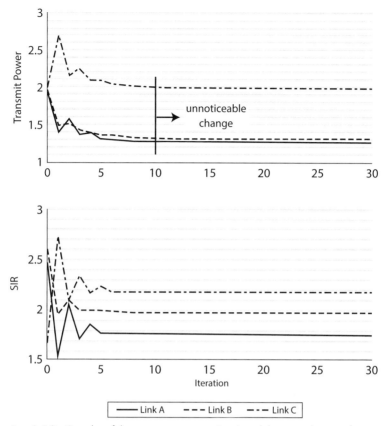

Illustration 1.18 Graphs of the transmit powers (top) and the signal-to-interference ratio (SIR) levels (bottom) for 30 iterations of the algorithm.

distributed algorithm (see illustration 1.19), as opposed to other, more **centralized** ones we will encounter (like Google's PageRank in chapter 5).

CDMA As a Standard

The DPC algorithm was proposed to deal with interference issues for CDMA. Even with this development, it took several large-scale demonstrations throughout the United States under realistic network conditions before major network operators would endorse it.

Finally, in 1993, CDMA was approved as a 2G cellular standard under IS-95, with the brand name cdmaOne. Three years later, the first large-scale commercial deployment of CDMA in the United States was made by

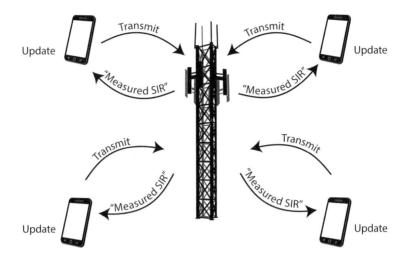

Update: "The ratio" x Current power

Illustration 1.19 Distributed power control is a fully distributed algorithm.

Sprint PCS. Though it has largely been upgraded to 3G standards, IS-95 and its immediate revisions are still used in parts of the world.

UPWARD AND ONWARD: 3G, 4G, AND BEYOND

The growth in the number of mobile subscriptions over the past few decades has been tremendous. The United States alone went from roughly 340,000 subscriptions in 1985 to 327,000,000 subscriptions 30 years later in 2015, an almost 1,000-fold increase. The mobile penetration rate in the United States has been larger than 100% since 2011.

Since the turn of the twenty-first century, 3G phones have gained momentum worldwide. The International Telecommunication Union's 3G specifications released in 2000 essentially called for cell phones to function as handheld computers: in addition to phone calls and texting, our phones are now capable of Internet access, video calls, and mobile TV. The two main 3G standards are UMTS, used primarily in Europe, Japan, and China, and CDMA2000, used especially in the United States and South Korea. Both technologies are based on CDMA and are typically deployed in the frequency range of 1.9–2.1 GHz.

Roughly 70% of the world's population was covered by at least one 3G network as of the beginning of 2015, up from 50% at the start of 2012. It is forecasted that more than four out of every five people (i.e., more than 80%) in the world will have access to 3G by 2020, making it almost ubiquitous. For information on the emergence of smartphones, check out Q1.5 on the book's website.

Since 1G networks were commercialized in the 1980s, a new generation of cellular network has emerged roughly every 10 years. Keeping on this track, the performance requirements for 4G were released in 2008. They specified higher speed requirements and capacities than the previous 3G specifications. Since then, the major standard that has emerged is long-term evolution, or **LTE**. Rather than using CDMA, LTE is based on a technology called **orthogonal frequency division multiplexing**, or OFDM.

The first LTE smartphones in the United States appeared in late 2011. At the beginning of 2015, roughly 25% of the world was covered by a 4G network, with this expected to increase to more than 60% by 2020. The predicted performance improvements over 3G are expected to attract 1 billion users by 2017. Though 4G network coverage worldwide was less than 3G as of 2016, it is also being deployed at a much more rapid pace.

The story of cellular evolution is a perfect example of how networks have struggled throughout the years to meet capacity demands of consumers. Different methods of sharing—frequency, time, and code-based wireless— have all been developed to do so. Even if we aren't aware of the processes involved, real-time updating and management of the power at which our calls are operating is essential to a cellular network's operation. Coming up with the right methods for sharing is difficult, yet very important.

Distributed power control illustrates several themes that are recurring in networking and in this book: negative feedback, system equilibrium, and distributed coordination. It also illustrates the following, major idea that we will see time and time again: allowing each user to make independent decisions driven by self-interest can be aggregated into a fair and efficient state across all users.

In the next chapter, we turn to WiFi, another type of wireless network. With WiFi comes a different flavor of sharing than cellular: rather than having stringent power control algorithms, WiFi relies on random access to manage interference among users in the same location.

2

Accessing Networks "Randomly"

By the mid-1990s, the second generation of cellular had gained momentum worldwide. The two competing technologies—TDMA and CDMA—each brought about much-needed improvements in network capacity.

Around this time, engineers began wondering whether they could think of a fundamentally different way to let the air be shared. Efforts in this direction led to the invention of **WiFi** technology.

TRAFFIC LIGHT VERSUS STOP SIGN

To start off, let's consider a simple analogy to car transportation to see how sharing in WiFi differs from sharing in cellular.

Suppose you're driving in your car, and you approach an intersection, like in illustration 2.1. If the intersection is governed by a traffic light, then while the light is green, the intersection is dedicated to your side. This is similar to how time, frequency, and/or codes are allocated to callers for the duration of their sessions in cellular.

As long as there's traffic volume on the other side, the red light makes sense. But what happens when you're stuck at the light and nobody is driving by the other way? That seems pretty wasteful (and quite frustrating). Why shouldn't you be able to drive through? In such situations, stop signs are a more efficient way to regulate traffic. When we approach one, we stop, look both ways, and then move on through if nobody is coming. We just need a form of coordination (stopping to look) to minimize the chance of an accident.

With the stop sign, rather than dedicating the intersection, we let everyone share it, as long as they follow the rules of random access. This is more efficient when there is a small and variable amount of traffic. Stop signs are similar to the way WiFi works: instead of having a dedicated resource, we

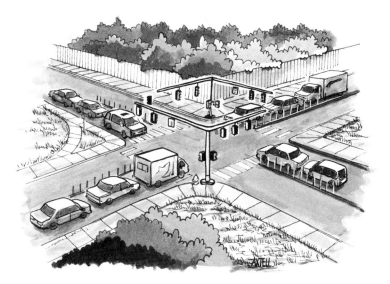

Illustration 2.1   The main differences between resource sharing in cellular and WiFi are similar to the differences between traffic lights and stop signs. Stop signs (WiFi) are more efficient when the traffic volume is small, but they are not scalable as the number of cars (devices) increases. In these situations, we prefer traffic lights (cellular), which regulate traffic flow by dedicating the intersection (the resource) to one side at a time.

have devices "listen" (i.e., "look both ways") before transmitting to prevent them from colliding with someone else.

But stop signs become problematic as traffic volume increases. When lines build up, the nature of "stop" and "go" one-by-one can cause extremely long waits, especially when the other side doesn't have a stop sign. In these situations, the dedicated nature of traffic lights may be preferred. As we will see, WiFi has a similar problem: its performance degrades drastically as the number of devices increases.

EMERGENCE AND EVOLUTION OF WIFI

What if we could create small stations to provide wireless Internet connection to people close by who don't move too fast? This idea forms the basis for WiFi, as you can see in illustration 2.2. It takes advantage of the benefits offered by random access.

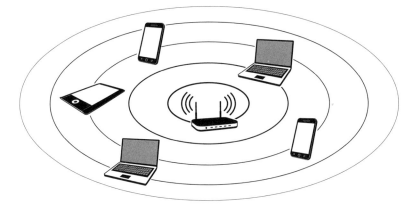

Illustration 2.2 WiFi provides wireless connections to devices (e.g., laptops, phones, and tablets) that are close to an access point.

In 1985, the FCC opened up some frequency bands to the public. Unlike the ones for cellular that companies must purchase expensive licenses to operate in, anyone can use these (as long as they follow some rules). Centered around 2.4 and 5.8 GHz, they are called the industrial, scientific, and medical, or ISM, frequency bands. Engineers seized the opportunity to use the ISM bands for communications, and what evolved into WiFi. They used them for other applications too; in fact, the most commonly encountered ISM device today is the microwave oven, because these frequencies happen to excite water molecules well.

Officially, WiFi is called **IEEE 802.11**, because of the scheme that the Institute of Electrical and Electronics Engineers uses to name their standards. The 802 part is for local area networks, echoing the fact that it is short-range, and the .11 part is for wireless. The more attractive name WiFi, short for "wireless fidelity," was coined and stuck.

With different groups developing WiFi technology, there needed to be a way to ensure interoperability among products claiming to use it. The WiFi Alliance was established in 1999 for this purpose. They stamp the WiFi logo on devices that conform to an IEEE 802.11 standard.

An Alphabet Soup

Like cellular, WiFi was improved significantly over a brief period. Various upgrades have increased customers' connection speeds substantially. How

Standard	Year	Frequency (GHz)	Maximum Rated Speed (Mbps)
-	1997	2.4	2
b	1999	2.4	11
a	1999	5	54
g	2003	2.4	54
n	2009	2.4 & 5	450
ac	2013	2.4 & 5	1300

Illustration 2.3 The progression of WiFi standards over time and some of their characteristics. There have been other variations not listed here.

is speed measured? As the number of bits per second, or **bps**. Today, WiFi speeds are usually referenced in the millions of bps, or **Mbps**.

The first WiFi standard was introduced in 1997; it offered a speed of 2 Mbps and operated in the frequency range around 2.4 GHz. To make the naming convention even more confusing, each WiFi upgrade adds a letter to the end of 802.11, but not in chronological order:

- In 1999, 802.11b was released, which uses the 2.4 GHz band and can transmit up to 11 Mbps. In the same year, 802.11a was introduced, transmitting up to 54 Mbps in the 5 GHz band.
- In 2003, 802.11g improved the 2.4 GHz top speed to 54 Mbps.
- The release of 802.11n in 2009 pushed the highest transmission rate to over 100 Mbps, operating in both the 2.4 GHz and 5 GHz bands.
- Most recently, 802.11ac was released in 2013; it is supposed to transmit at over 1 Gbps (1,000 Mbps) peak speed in the 5 GHz band.

You can see a summary of this progression in illustration 2.3. The maximum rated speeds are only attainable in theory, though. Under realistic conditions, you would be lucky to get more than a fraction of the advertised ideal speed.

With improving transmission speeds, the demand for WiFi services continues to increase. By 2011, more than a billion WiFi devices were being used around the world, with hundreds of millions being added each year. By 2014, this number had reached 4 billion worldwide, and is expected to hit 7 billion by the end of 2016.

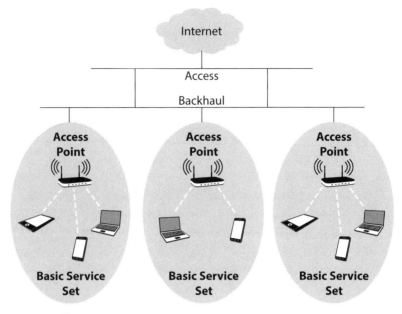

Illustration 2.4 Diagram of a typical WiFi deployment.

How WiFi Is Deployed

As mentioned, WiFi's method of "sharing the air" is very different from cellular's. But the way it's deployed is actually quite similar.

- Users in WiFi networks lie in what's called a **basic service set**, or BSS (or an **extended service set**), rather than a cell.
- Within each service set, users communicate directly with an **access point**, or AP, rather than a base station.

You can see the WiFi network layout in illustration 2.4. When your device searches for WiFi connectivity, it sends messages to discover which access points are in its transmission range. This results in a list of network names that you can choose from on your screen. Each of these user-friendly names identifies a service set, and they are more commonly known as **service set identifiers**, or SSIDs. You've probably been frustrated at one time or another from seeing a lock next to an SSID with good signal strength: if it's password protected, your device can associate with the access point only if you have the password to authenticate your status.

Each AP is typically connected to what is called the **backhaul**. The backhaul is a wireline deployment, often **Ethernet**, which is another (and much older) IEEE 802 family member. From this, the Ethernet connects to the access network, which ultimately provides a connection between the WiFi deployment and the rest of the Internet.

To send and receive over a WiFi network, a device needs to take care of a few things. It needs to select an AP in its range and choose the right frequency channel to use. While it's connected, it also needs to listen to the AP in setting its transmission rate, which can vary substantially depending on what the channel conditions are.

If you're interested in more information on these processes, check out Q2.1–Q2.3 on the book's website. There, you can also find more details about WiFi deployments (Q2.4–Q2.6). In the rest of this chapter, we are going to discuss one of the remaining tasks to be accomplished after the devices have gotten these other things right: managing interference. It's an essential part of sharing, WiFi's method being quite different than cellular's.

METHODS OF RANDOM ACCESS

When two transmitters that are within interference range of each other send at similar times, their signals will collide. More precisely, we speak of frames colliding, where a frame is a piece of digital data transmission (more on that in chapter 11).

There are three possible outcomes of a **collision** between frames. First is the worst case: both are lost. This means that the receivers of each frame could not correctly decipher them. Second is **capture**, meaning that the stronger of the two is received (see illustration 2.5). "Strength" here refers to SIR, the measure of quality that we talked about in chapter 1. Third is **double capture**, the best case: both are properly received.

So which outcome will prevail? Well, that depends very strongly on a few factors. We'll take the conservative approach and assume the worst case: whenever a collision happens, both frames are always lost.

We saw in chapter 1 how 3G cellular manages interference: give everyone different codes and use power control to adjust their volumes. Maybe we can just use that here? Not quite. WiFi has some properties that prevent power control from being an effective solution.

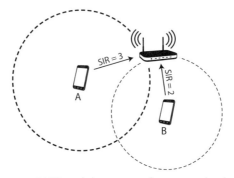

Illustration 2.5 When two WiFi mobile stations close to each other transmit a frame at around the same time, the frames will collide. Since A has a higher signal-to-interference ratio (SIR) than B, A's has a better chance of still being received.

For one, WiFi frequency bands are unlicensed: there's a lot more interference that we have no control over, since the spectrum is free for anyone to use. Also, it has a much smaller cell size: adding another person to a service set can make a big difference in interference conditions, because there are only a few to start with. Further, it has a smaller maximum transmission power: since the ISM band is unlicensed, we cannot crank the power up too high.

If we can't use power control, then how do we deal with interference? WiFi takes an entirely different approach: it tries to avoid collisions from happening in the first place.

Coordination Is Key

Think about TDMA (chapter 1), where each link has its own timeslot for transmitting. If we have three transmitters A, B, and C, the transmission sequence would go something like A, B, C, A, B, C, and so on. We are guaranteeing that each user will have the medium to herself for her timeslot each round. In our traffic analogy, this is like intersections run by traffic lights, where we only allow one side to come through at a time.

With WiFi, time is also the dimension of interest. Instead of dedicating timeslots, though, WiFi allows devices to transmit when they need to, as long as the channel appears idle to them. In other words, it requires them to remain cognizant of what others are doing in an effort to prevent collisions. This is like a driver coming to a stop sign: she has to look both ways to make sure no one is coming before proceeding.

Illustration 2.6 Revisiting the cocktail party analogy.

Both of these techniques are methods of **medium access control**. While TDMA is a form of dedicated access, WiFi is governed by **random access**.

We can think here again of the cocktail party analogy (see illustration 2.6), where we have guests' voices overlapping in the air. With enough interference, you can't understand what your friend is trying to say. TDMA is like arranging for the guests to talk at different times, by assigning them different times to speak. Random access separates the guests in time, but doesn't assign slots; rather, at a given time, a guest is allowed to talk as long as nobody else is.

With random access, each device is required to obey a certain procedure for deciding when to transmit and how long to transmit. The timeslots are shared by all the devices, because they can try to use the medium when they need it, as long as it's not currently occupied by another device. In other words, WiFi protocols require each device to obey some "courtesy procedure." The name for this procedure is **carrier sensing multiple access**, or CSMA, because each device tries to "sense" the presence of other devices in the channel.

Like DPC in chapter 1, CSMA is a fully distributed procedure. Each device executes it locally, using information it gathers through its own view of the channel. There's no need for some central coordinator to aid in the

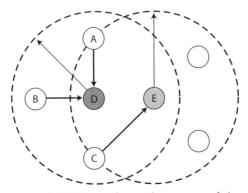

Illustration 2.7 An example WiFi topology. The ranges of the two access points, D and E, are indicated by the dashed circles.

decision making, another example of distributed coordination being a key theme of networking.

When the number of devices is small, WiFi can obtain really impressive bit rates. As more devices start competing for the same access point, however, the speeds will suffer substantially. Before diving into CSMA, let's consider a simpler random access protocol for which it is easier to quantify how the performance degrades with more devices.

ALOHA (FROM HAWAII)

Take a look at illustration 2.7. There are three transmitters, A, B, and C, and two access points, D and E. While A and B want to send to D and C wants to send to E, they are all in interfering range of one another.

At the beginning of a timeslot, each device is faced with the following question: should I send a frame or not? Some potential results of this "decision-making" process are given in illustration 2.8. If all the WiFi **stations** always answered "yes" to this question, then we would always have collisions. That's clearly an undesirable state to be in. A given station has to refrain from sending sometimes, to allow others to get their data through. In turn, it should expect that the others will refrain so that it can get its data through soon as well.

We could make whether or not a station will transmit dependent on a lot of different things (e.g., whether it has had a collision recently). What if we just assigned a fixed chance (probability) to it transmitting in each

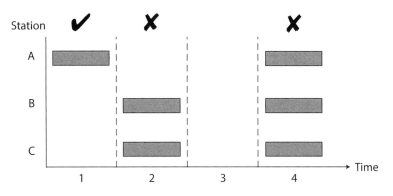

Illustration 2.8 Example of four transmission timeslots for A–C in illustration 2.7. The third slot is a ``wasted opportunity,'' because nobody transmits during this time.

timeslot? If 50%, then a station would send half the time, and hold back the other half. If 10%, then it will send one in every ten times on average, and hold back the other nine. This method is actually the basis behind the Additive Links On-line Hawaii Area, or **ALOHA**, protocol, invented in 1971 by Norman Abramson at the University of Hawaii (obviously, the acronym was not a coincidence!).

Illustration 2.9 shows what happens when we increase the chance of transmission in ALOHA. The higher it is, the more frames are sent, and the higher the chance of a collision. But if it's too low, we will have a lot of wasted opportunities. The question is, which choice will make the **throughput**—or rate of successful message delivery—as high as possible?

It All Comes Down to Throughput

In each timeslot, there are three possible outcomes: (i) successful transmission (i.e., exactly one station sends), (ii) collision (i.e., more than one station sends), or (iii) no transmission (i.e., no station sends). In modeling throughput, the first case is our interest.

For a successful transmission to happen, we need two things to occur. First is that someone transmits. For any given station, the chance of this happening is just the chance of transmitting.

Second is that the rest of the stations do not transmit. What is the chance of this? Well, the chance of a given station not transmitting is just the opposite of the chance that it does. For example, if there is a 40% chance of

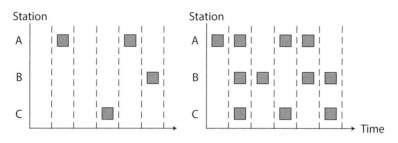

Illustration 2.9 This is what happens as the chance of transmission is increased.

transmitting, then there is a 60% chance of it not transmitting. Since we need all the others to not transmit, we multiply: the chance that two of the other stations don't transmit is $0.6 \times 0.6 = 0.36$, or 36%, the chance that three of the others don't is $0.6 \times 0.6 \times 0.6 = 0.216$, or 21.6%, and so on.

So, what's the chance that a station has a successful transmission? We multiply the chance that this station transmits by the chance that all the others do not transmit, that is,

$$\text{chance transmit} \times \text{chance no transmit} \times \cdots \times \text{chance no transmit}$$

For example, if A, B, and C are the only transmitting stations, and each has a 40% chance of transmitting, then the chance that, say, A is successful is $0.4 \times 0.6 \times 0.6 = 0.14$, or 14%.

This is a measure of ALOHA's per-station throughput. It's the chance that a particular station will have a successful transmission in any one times-lot. In our example, A is getting 14% of what the channel could provide with no interference.

To get the total throughput of the system, we add up the per-station throughputs. With the three stations A, B, and C considered above, it's $0.14 \times 3 = 0.42$, or 42%. Overall, the system is getting less than half of the throughput that could be achieved without interference.

To Transmit or Not to Transmit

We just saw two factors that affect the throughput in ALOHA:

- the chance that a station transmits in a timeslot, and
- the number of stations in the interference range.

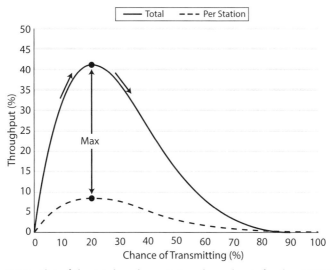

Illustration 2.10 Plot of the total and per-station throughputs for the ALOHA protocol for five stations.

Although it's hard to control when people come and go from our WiFi area, we can enforce the transmission chance as part of our "courtesy procedure." So how should we choose this value?

Let's fix the number of stations and see what the effect of changing the chance of transmitting is. Take a look back at illustration 2.7: which of A to E are stations? Actually, all of them: devices (A, B, and C) and APs (D and E) both need to send data. To make things easier, above we focused on just the devices.

With five stations, illustration 2.10 shows the total and per-station throughputs as the transmission chance changes. Starting from the left side, we see that increasing the chance initially makes the performance go up: with a small chance, there are few to no collisions, so we are just filling up the wasted timeslots and helping the throughput. Once the chance of transmitting hits 20%, the graph gets to its highest total throughput of about 40%. After that, the throughput drops: transmitting more is causing more collisions than it's helping.

For five stations, then, 20% is the optimal transmission chance, and the maximum throughput is 40%. This implies that the best we can do is about two of every five timeslots having a successful transmission, which is already not very efficient. What happens when we add more stations? That brings more potential interference, which we can counter by dropping the transmission chance more. It turns out that the maximum achievable throughput will stay about the same, slowly dropping to 37% as the number of stations gets really large. If you're interested in a detailed plot, check out Q2.7 on the book's website.

Overall, ALOHA doesn't scale well to a large number of users. This is a general theme of WiFi and is the price we pay for simple protocols.

PUTTING SENSING IN THE MIX

So, WiFi has many problems using ALOHA. Have they been fixed? Somewhat: the current WiFi protocols use CSMA, which gives better throughput.

ALOHA makes no attempt to coordinate the transmissions of the stations. Sure, it reduces the number of collisions by having people transmit less, and randomly at that. But what happens if I have a lot of data to send right now, and a timeslot opens up when nobody is sending anything? If my random "coin toss" tells me not to send, then I won't, and that's a wasted opportunity. Similarly, if the channel is currently occupied, I should be able to tell that, and refrain.

The problem is that ALOHA is purely random and does no sensing. In our stop sign analogy, that would be like putting a blindfold over your eyes and having you randomly choosing when to go without looking to see whether any cars were coming. Clearly, that would be extremely dangerous, risking the chance of a collision (though two frames of data colliding is less likely to be damaging than two cars colliding).

Sensing Carriers

Under CSMA, before a transmitter sends any frame, it will regularly listen to the air. This is called **carrier sensing**. You can see an example in illustration 2.11: A and C can hear that the channel is currently occupied and will refrain from sending anything until B is finished.

Once a station gets a glimpse of a quiet channel, is it allowed to begin sending? Not immediately. Rather, it has to first observe what is called a wait-and-listen period. If a station senses the channel is busy at any time during the wait-and-listen period, then it will stay silent, as does station B in illustration 2.12. This is analogous to what happens in typical conversations:

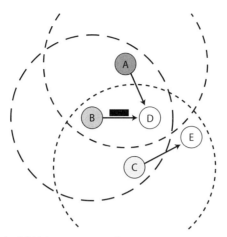

Illustration 2.11 With CSMA, stations perform carrier sensing, allowing them to see what is going on in their near vicinity. The dashed circles around the stations indicate how far they can hear.

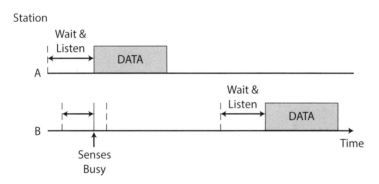

Illustration 2.12 Before sending, a station has to observe a full wait-and-listen period through which it does not hear any transmissions.

when you think someone has finished talking, it's common courtesy to give them a few seconds before you respond.

Once a station makes it to the end of a wait-and-listen period without hearing anything else, it can assume the channel is idle and start sending. This happens first for A, and later for B, in illustration 2.12. Obviously, even after a station waits, it's still possible that its frame will experience a collision. Perhaps another station began its waiting period at the same time as this one and decided to begin transmitting simultaneously.

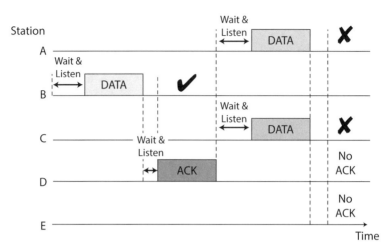

Illustration 2.13 Station B from illustration 2.11 needs to send. Sensing that the air is free, it goes through the wait-and-listen period, and, at the end, begins the transmission to D. Then D sends an acknowledgment message back to B. B receives this and knows everything went smoothly. After this, A and C don't hear anything, so they go through the wait-and-listen period and begin transmitting at the same time. Collision occurs, and they realize this when they don't get ACKs back.

How does a station know whether a frame it sent was properly received? Through feedback. With CSMA, when a frame is received, the receiver will send an acknowledgment, or ACK frame back to the source, notifying the sender that everything went OK. There is a wait-and-listen period before the transmission of an ACK frame, too. If the sender never sees an ACK from the receiver, it can assume that a collision occurred. You can see an example in illustration 2.13.

Patience Is a Virtue

What to do when a collision occurs? Each station needs to back off to a later time, at which point they will try to send their frames again. The question is, how does a station choose what time to back off to? Clearly, we don't want them to choose the same time: this would just cause another collision.

Instead, CSMA has each of the stations choose some random time in the future to retransmit. This point in the future is determined by the current **contention window** size. If a station's current contention window is 3, then it would choose some random number between 0 and 3. The number it

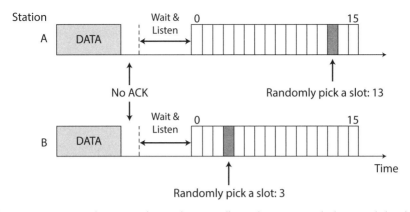

Illustration 2.14 When A and B realize a collision has occurred, they each back off based on their contention window size (15).

lands on tells it how many timeslots to wait before retransmitting (e.g., if it comes up with 2, it would wait 2 timeslots).

You can see this process in illustration 2.14, where two stations are shown having a collision. First, they go through the wait-and-listen period a second time. Then, they choose random numbers between 0 and the window size, 15, to determine when to transmit again.

The point of randomizing this selection is to reduce the chance that these stations collide again. Of course, it's possible that the stations will choose the same slot. Or maybe they will just collide with different transmitters altogether. If frames keep colliding, the interference condition in the air must be pretty bad. So all those stations experiencing persistent collisions of their frames should start backing off more aggressively. This is done by increasing the contention window size.

To help understand this, we can turn again to our cocktail party analogy. Suppose you're in a room having a chat with a group of people, and you have something you want to say. Someone is currently talking, and then they stop, which begins your internal "wait-and-listen" period. You give them maybe 2 seconds to keep talking if they want to, or for someone else to respond abruptly.

After that 2 seconds is up, you start talking. But someone else also tries to begin talking, and you collide. You're both caught off guard, and pause. You back off, wait 2 seconds, and then try to "retransmit." But she waits the same 2 seconds, and you again collide. This time you wait longer, and back

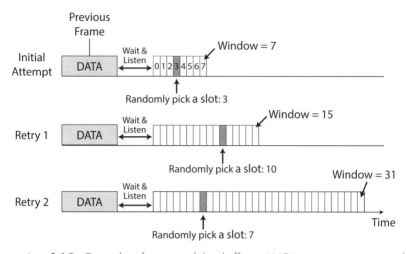

Illustration 2.15 Example of repeated backoffs in WiFi's carrier sensing multiple access. Collisions cause the contention window to double in length.

off further, say, 4 seconds. And so does she, causing another collision. Then, being polite, you decide to wait 8 seconds. Finally, she takes advantage of the lack of your voice and begins talking.

What was the problem here? Why didn't "backing off" work on the previous tries? Because you both waited the same amount of time, as luck would have it. Two seconds, then 4 seconds. Finally, in the third retry, you each get your message through, because you wait 8 seconds while your friend waits less. This is why randomization is important.

The progression 2, 4, 8, … is how the window size is increased with CSMA: by factors of two. Linearly increasing the window size (2, 3, 4, …) is one option, but people think it's not aggressive enough. So instead, CSMA mandates multiplicatively backing off. Since the multiplicative factor is 2, it is called a **binary exponential backoff**.

You can see an example of binary exponential backoff in illustration 2.15. To find the current window size, we multiply the previous number of slots by 2, and subtract 1. We subtract 1 because we want to include zero: it's possible that we won't back off at all and will transmit immediately. Initially, we have 8 potential backoff slots, 0–7, and so we say the window size is 7. Next, we have 16 potential backoff slots, 0–15, and so on. Multiplicative backoff will come up again when we discuss congestion control in chapter 13.

There are many more intricacies of CSMA that we won't have time to cover. For instance, it's possible for you to collide with someone you can't sense, which is known as the **hidden node problem**. This becomes an issue unless other precautionary measures are taken.

Comparison with ALOHA

When the parameters are tuned right, CSMA gives significant performance improvements over what ALOHA can achieve. For one thing, although the per-station throughput still decreases as we add more stations, it does so less rapidly than it does with ALOHA. Also, under CSMA, the total throughput will actually increase as we add the first few stations, whereas it is always decreasing for ALOHA.

But CSMA doesn't solve the collision problem altogether. In fact, after just a few stations are added, the total throughput will begin to drop in CSMA, too. Though it's an improvement over ALOHA, the rate at which it decreases as more are added is still substantial.

Overall, neither CSMA nor ALOHA is scalable to a large number of WiFi devices. This is why we see such poor performance at hotspots when there are a lot of people around. It's the price we pay for congestion in networks.

The focus of this part of the book up to now has been on different methods that have been developed to help us share network media (specifically, the air). But we have not said anything yet about the methods used to figure out what we should be charged for the resources we consume. Network pricing can also be an effective way of obtaining more efficient sharing. We turn to this topic next.

3

Pricing Data Smartly

Data charges make up a significant part of our cell phone bills. How do cellular providers set these price points? In this chapter, we see how so-called usage-based pricing schemes can send us better feedback signals than flat-rate, "buffet" schemes, leading to better sharing. Pricing can be a powerful way to manage networks.

PRICING LIKE A BUFFET

The data plans in our cellular contracts indicate how much we pay for the bytes we consume. First introduced by cellular providers as a way to charge for texting, these plans now encompass all the Internet applications we use on our phones, like surfing the web, streaming videos, and video chatting.

How are these plans structured? Think about how our utility bills—electric, water, gas, and so forth—are handled. They are typically based on the quantity of service that we consume. For example, if the electric company charges 10 cents for every killowatt-hour used, then someone using 500 kilowatt-hours will be billed $50, whereas if they cut their consumption in half, then the bill would only be $25. This "use more, pay more" type of pricing scheme sounds intuitive; is that how data plans are handled too?

Increasingly the answer is yes, but only in the past few years. Even though wireless cellular capacity is expensive to provide and difficult to crank up, consumers in some countries (e.g., the United States) used to have the luxury of only paying a fixed data cost each month no matter how much mobile data they consumed. Such a scheme is known as **flat-rate** pricing.

Are Buffets OK?

What do we mean by a flat-rate price? It's one that doesn't depend on how much you actually consume. Think of a restaurant offering a buffet (illustration 3.1): after paying some amount to enter, you can eat as much as you

Illustration 3.1 A restaurant buffet is based on flat-rate pricing.

want. It's in your best interest to follow suit: because you won't pay a penny more whether you have one, two, or five plates, you may as well eat what you can while you have the chance.

A buffet can be great for you, especially if you're hungry. But it may not be so great if you don't want to eat much, because then you pay the price of the buffet for just a small amount of food. And think about it from the restaurant's perspective: what would happen if people consumed larger volumes of food each time they came back? Could a restaurant keep offering buffets for the same price if customers kept doubling their appetites every year?

For years, data plans in the cellular industry were based on flat-rate pricing, while voice plans were not. The reason is that wireless operators in decades past saw voice calls and messages as the main use of cell phones, rather than mobile data. By paying some rate, say, $30 each month, the network would become a buffet for you, allowing you to consume as much data as you wanted. For a long time, such a scheme made sense for providers, because the amount of data consumed by cellular devices was low. The primary purpose of cell phones was voice, and data was viewed as a secondary, add-on feature.

As smartphones grew in popularity, this changed quickly. The demand for data began to rise rapidly as handhelds became capable of surfing the web, streaming music videos, and supporting a wide variety of other data-intensive applications. The introduction of the first-generation iPhone in 2007, for example, caused a 50-fold jump in cellular data demand. Many applications are also capable of running in the background (not requiring a human being to be operating them on the other side), and soon direct machine-to-machine and device-to-device communication will create even more data demand. Besides new applications, the sheer number of people using smartphones also fuels increases in demand.

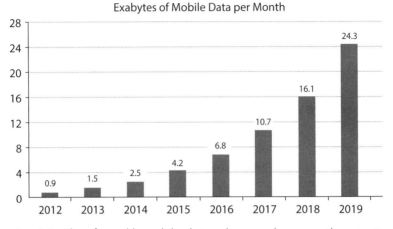

Illustration 3.2 Plot of monthly mobile data volume each year, with projections by Cisco out to 2019.

"Jobs's Inequality of Capacity"

Illustration 3.2 shows how high mobile data traffic had become by 2015, and how much it is expected to increase by 2019. The **byte** is the unit for measuring data size, but one byte is a rather small amount: an audio file typically consists of millions of bytes, and a video can be in the range of billions. So, we usually speak of data sizes in terms of megabytes (MB), or millions of bytes, and gigabytes (GB), or billions of bytes. In illustration 3.2, the numbers are in exabytes, with an exabyte being one billion gigabytes. That means in 2015, 4.2 billion GB of mobile data were already running through the Internet per month! The volume has increased, and is projected to continue increasing, about 50% each year.

Which data applications make up most of this consumption? Illustration 3.3 breaks down the percentage by application type for 2014. Video streaming, at 55%, accounts for the most consumption, followed by web browsing at 36%. Together, that is more than 90% of the data demand. With the exception of limited web browsing, these features were not available on mobile devices before smartphones and tablets.

Similar to how buffets only have finite amounts of food to serve their customers, networks only have limited capacities to support the flow of data to and from our devices. Soon after Steve Jobs introduced the first iPhone in 2007, a prediction was made that the growth in demand was beginning to outpace the growth in the amount of supply that could be provisioned

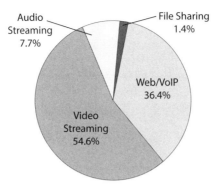

Illustration 3.3 Percentage of mobile data consumption by application type in 2014.

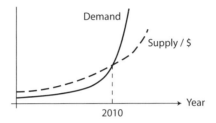

Illustration 3.4 The trends of demand and of supply per dollar of cost for mobile data capacity over time.

with each additional dollar spent on increasing the network capacity. The gap has kept widening in the years since.

You can see a sketch of this trend in illustration 3.4. We call it Jobs's inequality of capacity. Once engineers and developers figured out how to make it easy and attractive for users to consume mobile Internet data, user demand (and innovations in data applications) began to develop faster than the supply side could keep up with.

No technology can increase its cost-effectiveness this aggressively each year forever. A way of regulating the demand to keep it in line with the capacity was needed, so that the network could be shared more effectively.

USE MORE, PAY MORE

What was the solution that Internet service providers—or **ISPs**—came up with? Rather than charging a flat rate, they began switching to **usage-based** pricing schemes, in which customers are charged based on how much data

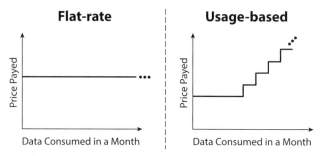

Illustration 3.5 Flat-rate and usage-based pricing send different feedback signals to consumers. With a usage-based scheme, the monthly cost increases for each bucket of data consumed above the quota (e.g., each additional GB), in a ``staircase'' fashion. Flat-rate, on the other hand, is completely independent of monthly consumption.

they consume each month. Charging this way sends a different pricing signal to consumers: instead of a flat rate each month no matter how much you use, they incur a cost for each bucket of gigabytes they consume. You can see this in illustration 3.5: flat rate is completely independent of consumption, while usage-based charges more for each batch of data consumed above some quota. The exact shape of the usage-based "staircase" depends on pricing details at a given time by a given operator.

This signal sent by the network serves as a form of negative feedback to the customer. In chapter 1, we saw negative feedback for power control: the network regulated each device's signal power based on the channel conditions by feeding back the measured SIR. The acknowledgments in WiFi's random access in chapter 2 are a form of negative feedback as well, which we'll look at more closely for internet congestion control in chapter 13. In the present chapter, the network is regulating each user's demand based on the available capacity by feeding back the usage-based price. Each consumer is thereby asked to internalize her negative externality (i.e., the congestion she creates) on the network by paying for the amount she consumes. Once again, feedback has come up as a general theme in network resource sharing.

Usage-Based Migration

There are two typical precursors to the introduction of usage-based pricing. First is that network usage has surged throughout the market, and that its demand is projected to climb faster than supply. For cellular, this happened with the introduction of iPhones, Android smartphones, and iPads. More widespread users and more capable devices create rising demand, causing users to consume more data. This incurs a higher cost to the ISP, because it must provision extra capacity. Usage-based pricing is a way to help the ISP's revenue catch up with the cost of supporting the rising demand.

The second precursor is that government regulations allow pricing practices to innovate. Although many regulatory issues are involved, allowing the monthly bill to be proportional to the amount of usage is among the least controversial.

By 2010, the migration to usage-based pricing in the United States had begun. In April of that year, AT&T announced its usage-based pricing for 3G data users. The following March, Verizon followed suit, first for its iPhone and iPad users and later in 2011 for all of its 3G data users. For the legacy consumers on unlimited data plans, in March 2012 AT&T announced that they would see their connection speeds throttled (i.e., slowed down) once their usage exceeded a certain amount. The 4G data plans from both AT&T and Verizon Wireless for the "new iPad" that launched soon after no longer offered any type of unlimited data options.

In June 2012, Verizon updated its cellular pricing plans again. A customer could now obtain voice and text at a flat rate, in exchange for turning their unlimited data plan to a usage-based one. AT&T followed with a similar move 1 month later. Similar measures have been pursued, or are being considered, in many other countries around the world for 3G, 4G, and even wired networks. Providers find it more important to have usage-based pricing for data than for voice, since the "primary" use of cell phones has now switched from voice to data.

Long Tail Getting Longer

Besides sheer capacity constraints, there are other reasons ISPs needed to migrate to usage-based pricing. An important one is that the heaviest data users are the ones whose demand also increases the most. If we group users into three classes based on their usage—light, average, and heavy—we'll end up with something that looks like illustration 3.6. As we move to the right on this graph, we have fewer users demanding that much capacity. But it is the rightmost, heaviest users who are often the dominant factor in how much it costs an ISP to manage a network. This tail has always been long, meaning that there are small numbers of users consuming very large

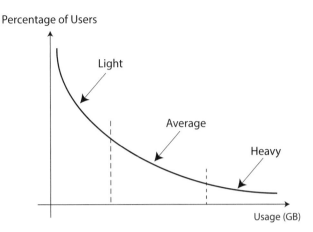

Percentage of Users

Light

Average

Heavy

Usage (GB)

Illustration 3.6 Some people are light users, some are average, and a small number of them are heavy users of data. The heavy users at the tail are the consumers that dictate the Internet service provider's cost structure.

amounts of data, but it's getting much longer now. As the tail grows longer, the mismatch between cost and revenue keeps increasing, unless a switch in pricing methodology is made.

Usage-Based Pricing Plans

What do the usage-based data plans offered by ISPs look like? In illustration 3.7, you can see the five plans that Verizon offered at the beginning of 2016. Generally speaking, there are three main characteristics:

- There's a baseline under which the charge is still flat rate. For instance, the middle Verizon plan will charge $60 for any monthly usage that does not exceed 6 GB.
- Above the baseline, the usage-based component kicks in. These Verizon plans charge $15 for each additional GB. The middle plan begins to incur the $15/GB charge when usage exceeds 6 GB. So for 7 GB, you would pay $75, for 8 GB, $90, and so on. The "staircase" means the pricing is done in increments of 1 GB.
- The overall charge is based on the total monthly usage. It does not matter when you use it, where you use it, or what you use it for, only how much you use.

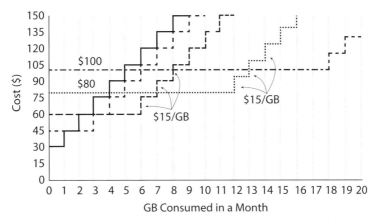

Illustration 3.7 Verizon's five data plan options in January 2016. Each line represents one of the plans that a customer can choose from.

Charging based on consumption sounds intuitive. After all, it's how most utilities and commodities are handled. But those used to flat-rate Internet connectivity may find it annoying at first. If content consumption is affected as a result, usage pricing will influence everyone on the food chain: consumers, network providers, content providers, app developers, device manufacturers, and advertisers. Yet we will see that there are several strong reasons, including those in the interests of consumers, that support usage-based pricing as a better alternative to flat-rate buffets.

SMART DATA PRICING

One way or another, the cost of building and operating a network must be paid by someone. Usage pricing based on monthly consumption, however, is not the only way that an ISP could tackle the issue.

Why not increase the flat rate for everyone? With a sufficiently high flat rate, the revenue collected would be adequate to maintain the network. However, this would be unfair to many of the users who do not consume much data and may not be able to afford the increase. Or what about capping heavy users' traffic? Once a consumer exceeds a threshold, he could be disabled from using the network. If you're interested in other possibilities, check out Q3.1 and Q3.2 on the book's website.

Alternatively, could we think of "smarter" versions of pricing altogether? Smart Data Pricing, or **SDP**, has been rapidly gaining momentum around the world since the mid-2010s. There are a variety of different SDP methods, which we can think of in three categories.

How to Charge

First, how should an ISP charge? As we've said, usage-based pricing is the norm now. In some countries, certain ISPs even reward customers for the unused portions of their mobile data quotas or allow them to trade those portions. The next step beyond this is congestion-dependent pricing, with special cases of time-dependent (i.e., charging less at times of the day with lower demands) and location-dependent (i.e., charging less at locations with lower demands) pricing. You can think of this as sending a more specific feedback signal to end users than what we have discussed so far for usage-based pricing: the price varies not only based on monthly consumption but also on the current congestion condition, and in turn regulates network demand and utilization at a finer scale.

For instance, consider Amazon's spot pricing on its cloud services: the price for the service fluctuates based on the current demand (and supply). As another example is the transportation system in London: higher prices are charged for public transportation in central business districts during weekdays.

Whom to Charge

Second, whom should an ISP charge? In addition to charging the direct consumers of mobile data, network operators may want to charge others on the food chain. What about the content providers that are getting the hits on their sites in the first place? In a **sponsored content** scheme, these providers may split the cost with the end users. Kindle eBooks followed this model. It also occurs sometimes with WiFi in airports, where you can use the Internet for free or for a relatively cheap price after watching some advertisement. Also, what about companies that permit a bring your own device to work policy for their employees? Through **split billing**, part of an employee's mobile data bill can be paid by their employers to compensate for the usage they accrue at work.

More generally, **zero-rating** or **toll-free** data is where people are charged less (or not at all) for data consumed for a specific application. Toll-free billing comes in two flavors: closed and open. Consider Facebook's

Illustration 3.8 If you're hungry, eating more slices of pizza will give you more ``happiness,'' or utility. The amount that you gain in happiness also drops with each additional slice.

`internet.org` initiative in 2015 to bring affordable Internet access to developing countries: this is a walled-garden model of closed toll-free billing, which is often viewed as incompatible with network neutrality. In contrast, a 1-800 number for mobile data is an example of open toll-free billing: anyone can sponsor the whole or part of the mobile data bill, not just those who go through a specific portal.

What to Charge For

Third, what should an ISP charge for? The basic approach, of course, is to charge for data usage. But why not also charge based on, for example, end user experience or Internet transaction? For example, some cloud providers also already charge based on a customer's desired quality of service, or QoS level, such as time it will take to complete computational tasks.

Through asking the questions of how, who, and what to charge for, SDP can achieve even more effective pricing signals, inducing even more efficient sharing. The days of unlimited mobile data in unlimited ways are gone: now we innovate around either "limited data" or "limited ways."

THE "TRAGEDY" OF FLAT RATE

We have already talked about some of the advantages of usage-based over flat-rate pricing. Among these was the fact that it sends a more effective pricing signal to the customers. By the end of this section, you will see why that is.

We'll first have to introduce some basic concepts in economics.

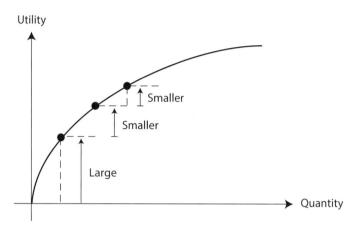

Illustration 3.9 Typical utility curve shape. More quantity always causes the utility to increase, but the change in welfare diminishes for larger amounts.

Higher Quantity, Higher Utility

Imagine there's a free pie of pizza near you (see illustration 3.8). You happen to be hungry, and it's a type of pizza that you like, so you want to get your hands on some slices.

You take your first slice. It tastes great, and certainly cuts down your appetite some. You want more, so you take your second. Again, it tastes quite good, and is satisfying, but nothing could match the satisfaction the first slice brought you: that was the one that started to satiate your appetite. After this, you're still kind of hungry, so you go ahead and eat a third slice. It still tastes good, but you're not as hungry anymore, so it's not doing you quite as much "good" as the first or second did.

This process keeps going (depending on how much pizza you can stomach), until eventually you are indifferent to eating another slice. As it's free, you might go for it since it tastes good, but you would not be willing to pay for another slice at this point.

This is an example of how a person's **utility** (i.e., "happiness") changes depending on the amount of resource allocated to them. Whether the resource be food, consumer electronics, cellular data, or something else, it is important to understand what a user's utility will be with the amount of the resource they get. As you can see in illustration 3.9, two characteristics are common to the way utility behaves:

- *Increasing*: It keeps increasing as the quantity allocated increases. With more data comes more benefit to you.
- *Diminishing marginal returns*: Above a certain point, it starts to increase more slowly. You probably put the first few gigabytes of data to the best use, and after that, the additional gain begins to drop off. This is called the principle of **diminishing marginal returns**.

Higher Price, Lower Demand

How would we go about quantifying someone's utility? One popular way is by observing how consumers behave with respect to the resource in question. We'll explore this method here, but if you're interested in other possibilities, check out Q3.3 on the book's website.

Whenever someone makes a purchase, that person is interested in making her **net utility** as high as possible. This is the person's payoff from the purchase, or the difference between her satisfaction and the price she paid. When the customer purchases an amount of something at a specific unit price (i.e., \$10/GB), the payoff is

$$\text{Payoff} = \text{Utility} - \text{Unit price} \times \text{Quantity}$$

The quantity that a user purchases in turn depends on the price charged, according to what's known as **demand**. As you can probably guess, a higher price induces a lower demand: for example, if the price of data doubled, you would consume less, whereas if it halved, you would consume more. The precise relationships are shown through demand curves, which can take complicated shapes in reality. For our purposes here, we are going to think about linear demand, as you can see in illustration 3.10.

Supposing the seller has fixed the usage-based price, it's relatively easy to find a user's utility and net utility, from the shapes in illustration 3.11:

- From the offered price, we can get the user's demand on the curve.
- The price that the user is charged is the unit price times the quantity. Geometrically, this is the area of rectangle B in the illustration.
- The user's utility is the total area to the left of the demand curve and below the quantity purchased. In other words, it is the combined area of the triangle A and the rectangle B: A + B.
- Finally, we can get the net utility by subtracting the price from the utility: (A + B) − B = A, the area of the triangle A.

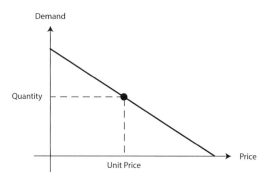

Illustration 3.10 A linear demand curve. It gives the amount of a resource a person is expected to purchase if the seller is offering it at a specific unit price.

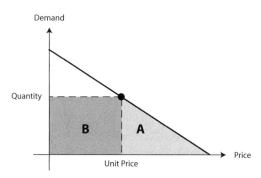

Illustration 3.11 Under a usage-based pricing scheme, the utility is **A** + **B**, the amount paid is **B**, and the net utility is **A**.

Consumption under Usage-Based Pricing

There is a more fundamental question to be answered regarding the demand curve: why is it that the quantity a user consumes will be fixed to this curve in the first place? In other words, why would a person not be incentivized to consume more, or to purchase less? The reason is that under usage-based pricing, the quantity at the unit price is the one that will make the user's net utility the highest.

To see this, let's think about what happens when someone lowers or raises her quantity consumed from the demand curve. The two cases are in illustration 3.12. On the left side, the user drops from the quantity on the demand curve: this causes the price charged to drop by the area **B1**, but it also causes the utility to drop by the entire area **A1** + **B1**. So, the net utility will decrease from what it was to begin with by the amount **A1**.

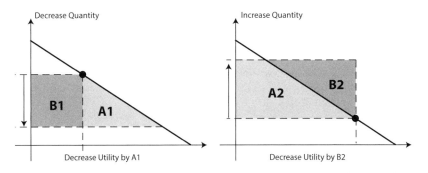

Illustration 3.12 Explanation of why it is in a user's best interest to stick to the demand curve under usage-based pricing.

The lessened utility outweighs the lessened cost. On the right side, the user raises his consumption from the original quantity: while the utility rises by A2, the price charged rises by A2 + B2, causing the net utility to drop by the area B2. The added cost outweighs the added utility.

This is why, under a usage-based scheme, it is always in the consumer's best interest to base consumption on the demand curve. Charging a unit price for data gives an ISP the ability to regulate user demand for data consumption over a mobile network. If the ISP sets the unit price based on the cost it incurs to operate and maintain the network, this will send an efficient feedback signal to consumers: one that forces them to internalize the negative externality they are imposing on the network from their consumption.

Consumption under Flat Rate

Does flat-rate pricing also entice users to stay on the demand curve? Remember that flat-rate pricing charges a single, fixed amount irrespective of the amount consumed. Under this scheme, then, it's in the best interest of a customer to grab as much as possible, until there is no more utility to be gained. The user should stray from the curve.

Think about it: if you pay $20 each month for your data plan, what would dissuade you from trying to stream 100 videos as opposed to 10? Clearly not money. You would consume as much data as you could possibly need each month, until you had no more use for it.

Illustration 3.13 In this analogy of ``the tragedy of the commons,'' a group of herdsmen are raising livestock on a shared parcel of land.

The Tragedy of the Commons

To see why flat-rate pricing is undesirable for the network, let's take a look at a well-cited analogy in economics called the **tragedy of the commons**, popularized by Garrett Hardin in 1968 (for more advantages of usage-based schemes, check out Q3.4 on the book's website). This analogy considers a group of herdsmen sharing a pasture that they are each using to feed their own livestock (see illustration 3.13). Each herdsman has the ability to add more cattle to graze on the land as he pleases.

Now, suppose one of the herdsmen, Bob, is trying to decide whether to put another animal in his herd. In an attempt to maximize his personal gain, he asks: "If I add one more animal, how will that affect my net utility?" On the one hand, if he adds another, he will receive all of the revenue associated with the sale of this animal. On the other hand, adding another brings the land one step closer to becoming overgrazed; once there are too many animals, the pasture will deplete, and all of them will perish. But this cost is spread evenly across all the herdsmen, so to Bob himself, it is only a small decrease in net utility relative to what he will gain from the addition.

Illustration 3.14 Driven by self-interest, each herdsman will keep putting additional cattle on the land. The tragedy of the commons results when the pasture becomes overcrowded, and eventually depletes, due to overgrazing.

Therefore, Bob decides that it's in his best interest to add another animal to his herd. Each time he goes through this thought process, he decides to add another. Every other herdsman faces the same decision, and does the same.

What is the end result? The pasture is only of a given size, so as more and more livestock are added, it will become overcrowded and overgrazed (illustration 3.14). Eventually, there will be no land left to graze. At this point, each animal will starve, which is the worst scenario for everyone. Without a proper pricing signal, Bob and the other herdsmen have collectively driven the pasture to ruin, and therein lies the "tragedy."

What do we make of this? The reason a person under a flat-rate data plan will keep consuming more is similar to the reason that each herdsman wants to keep adding more livestock. No additional cost is incurred each time the consumer draws from the network. From each person's perspective, the only "cost" is the minuscule congestion they are adding, so it's in their best interest to keep requesting as much as they can. But this could eventually lead to a "tragedy" where the network buckles under too much collective demand from everyone. It's a **negative network effect**.

The problem lies in the fact that we don't have an efficient pricing signal. Additional data translates to more utility, easily outweighing the extra "cost" that a person pays for putting more data in the network. We need to factor in that every data query is creating a negative externality on the network as a whole, like how our cell phones create interference for everyone as we saw in chapter 1. In this case, the way we make users internalize their negative externalities is by having them pay for each piece of data they request, or for each additional "cow" they want to add to the "pasture." With higher price comes lower demand, and then we can avoid the tragedy in the first place.

Summary of Part I

In this first part of the book, on the principle that "sharing is hard," we looked at two types of networks: cellular and WiFi. We use both of these daily, but often don't take much time to think about how so many people can share the air without ruining one another's conversations or causing capacity issues. We saw how the long-range, regulated cellular voice networks are privy to techniques, such as distributed power control through CDMA, where everyone adjusts their levels without causing an arms race. In contrast, we saw how the short-range, unregulated WiFi networks are more feasibly dealt with using random access techniques, where everyone tries to avoid collisions through sensing and backoff. Finally, looking at the case of mobile data, we saw how pricing can be an effective way of enabling more efficient sharing in a network. In each of these cases, negative feedback was important to making sure users had the right signals about network conditions and congestion.

In chapter 11, we will revisit sharing and pricing when we look at how the Internet is designed.

A Conversation with Dennis Strigl

Dennis Strigl is the former president and chief operating officer of Verizon Communications and the former president and CEO of Verizon Wireless.

Q: Thank you so much for taking the time to talk with us about wireless networks. We live in an age where we really breathe wireless networks each and every waking moment, and you were responsible for making a lot of those decisions over many decades in your leadership position across many companies, including as the CEO of Verizon Wireless. Over the past four decades of evolution in mobile communication, which part surprised you most?

Dennis: When you talk about four decades, you make me feel old, but okay. Let's start at the very beginning. I remember when I first came into the wireless business, we had an estimate from a consultant, McKinsey & Company, that said by the turn of the century—2000—there would be 900,000 wireless subscribers. We beat that hundredfold easily. And if you think about worldwide subscribers today, there are 344 million as of the end of 2014, so the growth of the industry has far exceeded the early estimates by McKinsey, Bell Labs, and AT&T. And we can attribute that to an outstanding network, a good product that works for our customers, and also to the fact that prices have continually come down over the years, from what was an executive's backseat-of-a-limousine device to what is now an everyday device, with penetration rates (meaning the number of people who use them around the world) at over 100 percent, about 110 percent as of the end of 2014—meaning that many people have more than one device.

In fact, I remember very early on, when one of my bosses said to me, "go run the cellular company, I know nobody wants to do it. I'll promise you that I'll only keep you there a couple of years, and then we'll bring you back to run one of our great telephone companies." And at the time I thought, I'll do whatever I'm told to do here. But it was sort of a "pat on the back and do us a favor." Just "go run this thing for a while," because it's going nowhere.

Q: The scale of market penetration over the past number of decades was the most surprising part?

Dennis: Not only the scale in terms of number of customers but also in terms of what people use devices for. Certainly voice messaging, and voice calls, the original design of the wireless network, grew exponentially over many years. But then as data was introduced, data too grew exponentially. As of [2008 or 2009], over a trillion text messages are sent a year. By the way, I think half of those my kids themselves generated. But in any event, I remember text messaging back in the late 1990s, and my network staff came to me and said, "take a look at this. We can put words over the cell phone." And I said, "who would ever want to do that?" And, of course, now we have pictures, we have streaming video, and it has grown into a phenomenal industry, one that has created many, many, many jobs around the world, and continues to do so.

Q: Yes, both the scale and the heterogeneity of the usage are surprising. What do you think supported this scale and heterogeneity?

Dennis: I think primarily two things: the first is the quality of the network and its improvement over time. As more cell towers were built, and as more radio spectrum became available, the quality of the service improved. By the way, customers also demanded better service. We knew that we weren't going to be able to grow our customer base significantly if the service remained the way it was in 1984, 1990, or, for that matter, 1995, so we had to incrementally improve the service over the years.

The second part comes from pricing. Originally a wireless phone, the device itself, cost about $3,000. There was a transceiver that was placed in the trunk of the car to make it work, and the cost of the service itself was tremendously high. It was $50 a month plus 40 to 50 cents a minute, depending on where you went in the country, and that included calls that weren't answered or calls that were dropped. If you look at what happened when competition came full-bore into the industry, which was probably in the mid-1990s by the time we had three, four, or five competitors, prices came down to a small percent of what it had been when wireless was introduced.

Q: When you were the CEO of Verizon Wireless, in the process of rolling out 3G and deciding on 4G, there were a lot of different competing proposals on the table. What was the most challenging decision that you had to think through and make during the 3G/4G revolution?

Dennis: First and foremost, the challenge was, how do you devote a block of your spectrum and a block of your system to voice and a block to data? How do you get that ratio correct? Initially, of course, there was much more voice than there was data, so that our engineers continually had to make sure that as they installed data capability in the network, both in 3G and in 4G—but a little bit more complex a problem in 4G—that it was in sync with the customers who had devices that would use that data, part of the network. So that was the big engineering challenge.

The big financial challenge that we had was that we spent almost $15 billion a year on buildings, plant and equipment. Most of it went to the network—cell towers, switching machines. And the challenge there, too, was, what do you spend it on? Do you spend it on the data portion or do you spend it on the voice portion, and can you possibly continue to keep spending $14, $15, $16 billion a year? Does the revenue justify that? And what was occurring at the time—and this is the real financial challenge— was that as data usage increased, voice usage leveled off, so that in the early 2000s, what we saw was a diminishing portion of our revenues coming from voice and a growing portion coming from data. But at the time, most of the data was relatively inexpensive for our customers.

Q: There were companies pushing for WiMAX and those pushing for LTE, and now it's quite clear that LTE is widely deployed globally. And was Verizon Wireless the first major operator to deploy LTE?

Dennis: Yes. To deploy long-term evolution, we were the first, yes.

Q: But ten years ago it wasn't clear where this was heading. Was there some interesting story you have about that decision?

Dennis: Well, there was always the battle within the industry about which technology would survive. WiMAX versus LTE is analogous to GSM [global system for mobile communications; see chapter 1] versus CDMA for voice. It was the same principle. And, of course, we believed that CDMA technology was always more cost-efficient, and we also believed that LTE technology was more cost-efficient. By the way, not to knock WiMAX—I think it's a good service—but it certainly didn't provide the speed at the same cost efficiencies that one could find from LTE.

Q: Back to pricing voice versus data. A mere few years ago, say, in the United States, one would typically get unlimited mobile data and only a limited number of voice minutes and text messages. Now it's increasingly hard to find unlimited *mobile data. I recently walked into a Verizon store and I found this poster saying small/medium/large/extra-large sizes as family-shared data plans with roughly $10 a gigabyte. Many other operators around the world, when they switch to LTE, give you usage-based pricing. What caused this change?*

Dennis: Simple—supply and demand. The demand had been on minutes of use. You were using your voice minutes. You were not using your data minutes, because they weren't there. Either the applications weren't there, weren't built yet, or you hadn't—we hadn't—figured out how convenient this was. So in the mid-2005/2006 years to the present, we noticed the minutes of voice use had leveled off. Well, now we're seeing the minutes of voice usage decline and the usage of data—whether it's gigabits, megabits, however you want to measure it—accelerating tremendously, exponentially. And so the question becomes, how does the user and the carrier pay for this huge investment they have in their network? But if you look at what you pay currently for 4G per megabit, it has come down substantially from what you paid under 3G.

Q: People have also started exploring a variety of smarter ways to price data, including, for example, dynamically pricing them based on network condition, or sponsored data that comes in different forms. There are companies rolling out an open toll-free service or what they call a 1-800 number for mobile data. On these and other ideas for matching the demand and supply for mobile data, where do you see pricing evolution going?

Dennis: Well, let me back up for just a minute and say that usually when you have free access, it comes with some kind of catch to it. Often you watch an advertisement or you watch continued pop-up ads that may annoy you and disturb your train of thought, so that's kind of the balancing act.

So where do we go with this? One thing I can almost guarantee is that the price per unit of usage will continue to come down as the usage grows, the consumer and the commercial usage grows and the revenue stream to the carrier grows, and as the costs continually come down both from the manufacturers and from the carriers themselves.

Q: There is also a lot of talk these days, and you see this even on primetime commercials, about the Internet of Things or Internet of Me and Internet of Everything. We're talking about networked devices that are inside our body, on our body, around our body, or in cities or industrial and agricultural settings, and so on. Where might we find the first couple of major industry sectors where it would take off and bring the most direct value?

Dennis: I completely agree with you on the Internet of Things. As many have said, we have tiny machines that control our lives in one way, shape, or form. In almost every way possible, whether they control our cities and usage of power and light, for example, and water resources, utilities, or whether we see it in transportation—which is probably, to your question, the first place that we will see it, so not necessarily just smart cars but smart highways, and you're already beginning to see that.

One of the reasons why it grows within verticals as quickly as it does is because not only is it convenient, not only is it fast, but it brings down the costs for the people providing it, so I think we'll see a lot more of that. And I don't think it's limited to specific verticals.

Q: Speaking of, say, transportation, we see the cloud descending to be among the end users or near-user edge devices, forming a "fog network." People look at this and are wondering about what kind of applications can now be enabled because of the physical proximity of these network devices. At the same time, people also are concerned about security and privacy issues. If you look at the competing interests from different parts of the food chain, everybody wants to be the owner. The Apples and Samsungs want to say the phone is the hub of this fog network, and the Ciscos of the world want to say the dashboard that they're working on with the automobile companies is the hub.

Dennis: Or the router or the switch or ….

Q: Or Verizon would say that that little base station that I control next to you, that's the hub. So what would be a good strategy for the network operator, and where might the hub end up staying?

Dennis: Well, my counterparts in the telecommunications industry aren't going to like my saying this, but I think at the current course and speed, it's going to end up on your device that's in your pocket, in your purse, next to you in your car. You don't leave home without it anymore. It's on you almost 24/7. So what do the carriers do to compete with this? I'm not sure. But I think the answer for the carriers lies in partnerships, whether it's partnerships with Apple, with Google, with the likes of Cisco, even manufacturers—Ericsson, Nokia. Those who own the customer and those who own the technology on devices, they must come together.

Here's the great advantage that Verizon has today. They have over 100 million customers. Now, Google may be able to say that in terms of their search service, but they can't say it in terms of the customer. Now, I have an Apple phone, but my Apple phone is connected to the Verizon Wireless network. Who bills me for this? Apple one time perhaps, but Verizon Wireless every single month. So both sides of this equation, of this competing—in some way, shape, or form—for the customer, they come together in partnership. We're at the very beginning of this.

Q: I have one last question, which is related to national policy, perhaps, and driven by all these applications and demand that we just talked about. The FCC declared that the United States was officially in a spectrum deficit as of 2014. Have we run out of spectrum? And if so, how do we live in the future of a spectrum crunch?

Dennis: I don't believe a word of this. There is not a shortage of spectrum either in the United States or around the world. The fact of the matter is that there is more spectrum than companies in the US can ever use. The challenge is that it is in the wrong hands. It is in the hands of the government, of agencies within the government that are unwilling to give it up under any circumstances whatsoever. The spectrum that they have is used terribly inefficiently, and what the government must come to realize is, they have to deploy the same efficiencies to their use of spectrum that commercial industry deploys. If they were to do that, they would not see a spectrum shortage.

Q: So you think the heart of the problem is efficiency of usage, rather than a shortage of spectrum itself?

Dennis: There are many in the government who would argue with me over this, but I have held this position for years, and the amount of spectrum that has been auctioned off for commercial use is insignificant compared to the amount of spectrum that is already in use. By the way, the broadcasters also are not efficient users of spectrum, so therein lies the issue of efficiency.

Q: Thank you very much, Denny. It has been a most illuminating conversation, and thank you for sharing your perspective. No matter which way it goes, it's going to continue to be interesting.

PART II
RANKING IS HARD

In part I, we saw how challenging it can be to efficiently divide shared resources among many users. Now, in part II, we turn our attention to another challenge in a network: how to find an appropriate ranking of a set of items. We focus specifically on the ranking methods used by Google, which is constantly updating the orders in which webpages are displayed to each of its site's visitors. In fact, Google has two different "types" of pages: those that occupy its ad spaces, and those displayed as part of its search results. We talk about these two topics in chapters 4 and 5, respectively. As you will see, very different techniques can be used for ranking items, with ad spaces being allocated to buyers based on bids, and search results being ordered based on measures of importance and relevance.

4

Bidding for Ad Spaces

If you've been on the Internet in the past decade, you've used Google search (see illustration 4.1). We go so far as to "google" everything from the closest coffee shops to the detailed answers for our college coursework. "Googling" was added as a word to the Oxford English Dictionary in 2006. A large portion of readers perhaps will be referred to this book by Google in the first place, too.

How does Google rank webpages? Even before this, you have probably considered something else just as important. With more than 55,000 employees, it generated revenue of $66 billion in 2014, up 19% from the previous year. So how does Google make money from its advertising business?

THE ONLINE AD BUSINESS

Magna Global estimated that the revenue generated by the online ad industry worldwide surpassed $160 billion in 2015. Google, in particular, generated almost 90% of its revenue from ads alone in 2014, leaving only 10% from other sources like Google Play, Google Apps, and Google Fiber, as you can see in illustration 4.2. How do these ads work?

Pay per What?

The origin of online advertising traces back to the early days of the Internet (the mid-1990s). Banners displaying ads were originally sold by websites (i.e., the sellers) to advertisers (i.e., the buyers) on a **pay-per-thousand-impressions** basis. This means that each time the buyer's banner aggregates 1,000 new views, the advertiser will have to pay the seller some predetermined amount of money.

Is this really the best way to charge advertisers? Naturally, seeing an ad does not guarantee that a person will click on it, let alone buy something

Illustration 4.1 Google's ubiquitous, trademarked logo.

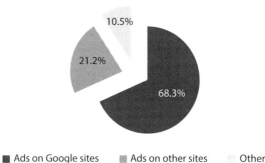

■ Ads on Google sites   ■ Ads on other sites   Other

Illustration 4.2 Google's approximate revenue by source in 2014. About 68% came from advertisements on Google's own sites, and 21% from those on partner sites. Only 11% came from sources outside advertising, though these sources are making up a larger percentage each year.

from it. So this method is a limited indication as to how much benefit the advertiser will gain from putting the ad up.

Then what about paying by *click*, rather than by view? In 1998, the search engine company GoTo began offering this alternative. In this model, advertisers would submit bids for how much they were willing to pay to appear at the top of the search results page (in response to specific search queries). These bids were on a **pay-per-click** basis: the advertiser would pay the amount of their bid to GoTo every time a person clicked on the link from GoTo's page to their website. The list of search results provided by GoTo was in descending order of the pay-per-click bid amounts, as you can see in illustration 4.3.

GoTo was widely credited as having created the so-called sponsor search model, also known as **search ads**. Overture (to which GoTo was renamed in 2001) was acquired by Yahoo! in 2003 for their own search marketing services. By the turn of the century, another search engine had begun using pay-per-click advertising widely as well: Google. Founded in 1998, Google had established its AdWords division to head the company's advertising by

Illustration 4.3 A small example of a search results page. The query is "Food stores," so the links that show up are related to food shopping. The ads are arranged by how much the advertisers bid for the first slot. Ben's Bakery offered the most, at $5, followed by Matt's Market ($3) and Gabby's Grocery ($1). The stores will then pay $5, $3, and $1, respectively, whenever someone clicks on their links.

2000. At this time, AdWords began offering keyword advertising with pay-per-click pricing as a feature of Google's search engine.

Since then, Google has become by far the largest search ad provider. According to eMarketer, Google possessed 55% of the market in 2014, with Baidu, the second-largest company in the business, at under 8%. Let's take a closer look at AdWords now.

Advertising Online

Suppose you want to advertise your site through Google search. Through AdWords, you would input the content to be displayed, including the link to your site and some descriptive text, and assign some keywords to your ad. This information will then be sent to Google's database, and your ad will be created: when someone enters a query into the Google search bar containing words that have been associated with your ad, yours may pop up in the results.

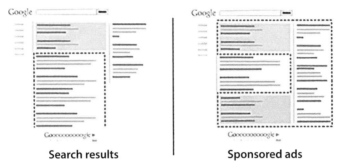

Search results   **Sponsored ads**

Illustration 4.4 On Google (in 2015), the standard search results (left) are typically shown in the middle of the page, while the sponsored ads (right) are shown along the edge of the page.

When a keyword (or a series of keywords) is entered into the Google search bar, you've probably noticed that Google actually returns two things: a standard list of search results, and a list of ads that have keywords matching the query. You can see a typical search layout (as of 2015) containing these two types of results in illustration 4.4. The standard search results are shown in the middle of the page, while the sponsored ads are shown along the edge of the page: on the top, right panel, or bottom.

Try typing something into the Google search field. As a sample, we entered "Online Education" (which, by the way, we'll discuss in chapter 8) in June 2015. Aside from the 10–15 standard search results that had links to certain online programs and rankings, there were ads at the top and on the right panel that Google returned based on the keywords. The first one at the top was for the University of Phoenix (`www.phoenix.edu/ Education?`). This space and the one at the top of the right panel probably receive the most clicks, and are the ideal locations for your ad to appear.

Where will your ad be placed in the list? This answer depends on how much you are willing to pay for sponsorship, and is allocated through an auction. In this auction, you and the others aiming for these keywords are the bidders. At the end of the auction, the highest bidder will get the best spot, the second highest the second best, and so on. We look at different auction mechanisms throughout this chapter.

When do the advertisers pay Google? With pay-per-click, Google gets paid every time an ad is clicked. The average number of times an ad space

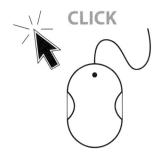

Illustration 4.5 The click-through rate of an ad space, measuring the expected number of visits in a given amount of time, is an important consideration for an advertiser when submitting a bid.

is clicked by a viewer, say, over 1 hour, is called the **click-through rate**. The payment that an advertiser makes to Google is proportional to the click-through rate of the ad, because this rate is an indication of the value Google is providing (see illustration 4.5).

What's in it for the advertisers? Each can expect some average revenue, in dollars, from each click on their ad. The product of the click-through rate (clicks per hour) and the revenue per click gives their expected revenue for a particular ad space, which is the **valuation** of the ad space to the buyer. Each advertiser will have a different valuation for each space.

For example, let's say an ad space receives 20 clicks each hour, and for each click, there's a 50% chance that the viewer will purchase an item. What is the valuation of this space to an advertiser whose average cost of an item purchased is $70? The expected revenue is

$$20 \, \frac{\text{clicks}}{\text{hour}} \times 0.5 \, \frac{\text{purchases}}{\text{click}} \times 70 \, \frac{\text{dollars}}{\text{purchase}} = 700 \, \frac{\text{dollars}}{\text{hour}}$$

OPEN AUCTIONS

As we said, Google sells ad spaces to buyers through ad **auctions**. In general, an auction has a number of bidders, items, and sellers, but we focus on the case of a single seller, because Google is the single seller here.

Though auctions gained much of their popularity with the rise of the Internet, they were used as a means to negotiate the exchange of commodities for many centuries before the days of Google. Items ranging from horses and livestock to real estate and entire empires have been placed on

Illustration 4.6 The classic, open, public auction room.

the auction block throughout history, with the earliest auctions dating back as early as 500 BC. As a (relatively recent) example, during the American Civil War (1861–1865), goods seized by armies were sold via auction by the local commanding colonel; this is why some auctioneers in the United States today are called "colonels."

When you picture an auction, you likely think of a public venue with a single auctioneer and multiple bidders, as in illustration 4.6. The auctioneer stands in front of the room next to the item being auctioned, mediating the process, with the bidders in the crowd.

This type of auction, in which everyone's bids are publicly announced, is called an **open auction**. There are two main types of open auctions: ascending and descending price. In an **ascending price auction**, the auctioneer first announces a base price, and then any of the bidders can raise their hands to bid to a higher price. He/she may say "Bidding starts at ten dollars! Any takers for ten dollars?" Then someone raises their hand: "We have ten dollars, anyone going for more than ten dollars?" Then someone offers twenty dollars, and so on, until nobody will increase more: "One hundred going once, going twice, gone!" This last bidder wins the item and pays his/her most recent bid.

The ascending price auction is the type we are most used to. With a **descending price auction**, in contrast, the auctioneer announces a high price first, so high that nobody will accept it. Then he/she gradually lowers

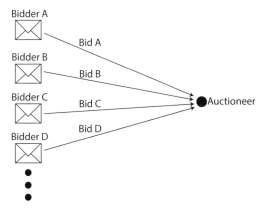

Illustration 4.7 In a sealed-envelope auction, each of the bidders (A, B, C, ...) submits a bid privately to an auctioneer. The outcome of the auction is clear: for a single item, the highest bidder will win it. Determining what to charge is more tricky.

it down until there is a bidder that shouts "OK!" This bidder wins the item, and pays the price at which the bidding stopped.

Open auctions are used quite frequently, but not by Google and other search ad sponsors. The alternative to a public venue is private disclosure of bids, called **sealed-envelope** auctions, which are much more practical in many settings. We take a closer look at them next.

SEALED-ENVELOPE AUCTIONS

Google uses a specific type of sealed-envelope auction to allocate ad spaces to bidders. Before considering multiple items (e.g., multiple ad spaces), let's first take a look at single item auctions.

In a sealed-envelope auction, each bidder will submit her bid privately, and all the bids are revealed simultaneously to the auctioneer (see illustration 4.7). The auctioneer then decides the **matching** (i.e., allocation) and how much to charge for it. The matching part is easy: the highest bidder gets the item. But the amount charged will vary, depending on what type of sealed-envelope auction it is:

- In a **first-price auction**, the winner pays the highest bid, which is her own.
- In a **second-price auction**, the winner pays the second-highest bid.

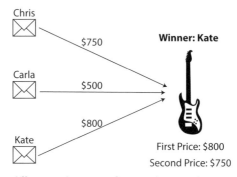

Illustration 4.8 The difference between first- and second-price sealed-envelope auctions is in how much the winner pays for them.

You can see the difference between these two in the example in illustration 4.8. Kate is the winner of the guitar, with the highest bid of $800. Under first price, she would pay her bid of $800, and under second price, she would pay the second-highest bid (Chris's) of $750.

Why would we ever use a second-price auction? Intuitively, it sounds wrong. If I want an item, and I know the winner will pay the second-highest bid, why not bid an extremely high amount, way above my own intrinsic valuation of the item?

Well, it turns out that this intuition is actually the part that is wrong, for a simple reason: if everyone engaged in this same strategic thinking, then the "second-highest bid" would also be extremely high. So the winner would end up paying much more for the item than she thinks it's worth. This knowledge should discourage everyone from bidding higher than their true valuations, to avoid paying more than what it's worth to them. We'll take a look at this in more detail next.

The Second Price Is Right

At the end of the day, a bidder's goal in an auction is to maximize her **payoff**: the net benefit she receives. If a bidder wins an item, her payoff is the difference between her valuation and the price she pays:

$$\text{Payoff} = \text{Valuation} - \text{Price paid}$$

However, if she loses, her payoff is zero. Payoff is actually a special case of net utility, which we discussed in chapter 3 (see page 55).

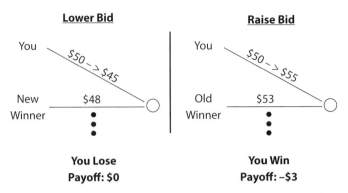

Illustration 4.9 In a sealed-envelope, second-price auction, it is in your best interest to bid truthfully. Raising or lowering your bid will only reduce your payoff.

Can the payoff be negative? Absolutely. The price paid may very well be larger than the valuation. This is a scenario that a bidder always wants to avoid, because in this case she would rather not receive the item at all.

How can a bidder maximize her own payoff? That answer depends on the type of auction being used. Consider first the case of a sealed-envelope, first-price auction. What is the winner's payoff in this case? Because the price she will pay for the item is her own bid, her payoff is her valuation of the item less what she had bid. In general, it's difficult for a person to determine what she should bid to maximize her payoff, because whether or not she wins depends on everyone else's strategy. But one thing is clear: in this type of auction, everyone should always bid below their individual valuations, because otherwise, winning the item can only result in a non-positive payoff (i.e., either zero or less than zero).

How about in a second-price auction? Here, the winner will pay the second-highest bid, so her payoff is her valuation less this number. This type of auction encourages **truthful bidding**, meaning that the best strategy of each person is to submit her true valuation as her bid.

How come? Because changing the bid cannot improve the payoff. The reasoning is similar to why usage-based pricing encourages users to consume based on the demand curve, as we saw in chapter 3.

To see this, let's say you are bidding on an item for which you have a true valuation of $50. You are thinking of submitting your true valuation as your bid, but you also have two other options: either raising or lowering your bid, as shown in illustration 4.9. First, what would happen to your payoff if you decided to lower your bid to, say, $45? In a second-price auction, the only way this would change the outcome is if you were originally going to win, and you lower it enough that the second-highest bidder now wins. Maybe this person had submitted $48. Originally, you had a positive payoff: you were going to pay the second-highest bid of $48, and your valuation was $50 − $48 = $2. But now that you lose the auction, your payoff is zero! You should have bid your true valuation, because positive payoff is better than zero payoff.

On the other hand, what would happen if you raised your bid to, say, $55? The only way this would change the outcome is if you were originally going to lose the auction, but you raise it enough that you now win, causing the previously highest bidder to lose. Let's say this person had submitted $53. Originally, you had zero payoff. Now that you win the item, you pay $53, meaning your payoff is $50−$53 = −$3. This value is negative, because your valuation is lower than the second-highest bid. You should have bid your true valuation here, too, because zero payoff is better than negative payoff.

So a second-price auction is "better" than a first-price one, in the sense that it's in everyone's best interest to bid truthfully. As bidders think about what their bids should be, there's an implicit feedback signal that encourages them to set it to their intrinsic valuations. It is useful to decouple the decisions of who wins and what price the winner pays.

Is second price better than, say, a third-, fourth-, or fifth-price auction for that matter? It turns out that second price is the only one that will make each user capture her own negative externality that she imposes on the network: if we took the winner out of the auction, then the second-highest bidder would have won instead. In this way, the winner is paying for the fact that the second-highest bidder did not win, because the winner's payment is the second-highest bidder's bid.

As with distributed power control in chapter 1 and usage-based data pricing in chapter 3, the second-price auction is yet another example of how forcing users to internalize their negative externalities by sending them signals about their impact is a common theme in networking.

Some Unexpected Connections

In illustration 4.10, you can see a taxonomy of the single item auctions we have discussed. There are some interesting similarities between them.

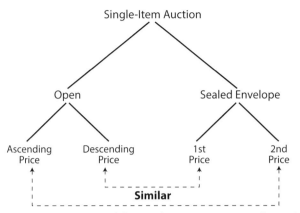

Illustration 4.10 A taxonomy of the single item auctions we have discussed.

Think first of an ascending price, open auction. While the price is increasing, all the bidders have their valuations in mind, and each will stay in the game until the current bid is higher than his/her valuation. The current price will keep escalating until the bidder with the second-highest valuation opts out. So, unless the winner increased the current price drastically, he/she pays the second-highest bidder's valuation (plus some small difference). In this way, ascending price is similar to second-price auctioning.

Now, consider a descending price, open auction. The auctioneer will keep decreasing the price until it reaches the highest bidder's level, at which point the highest bidder will stop the auction. As long as this bidder doesn't wait for it to drop further (i.e., she is cautious), she will pay her own valuation, which is the highest price. In this way, descending price is equivalent to first-price auctioning.

Some single item auctions are neither strictly closed nor strictly open. For example, in eBay auctions, bidders are given some idea of what the highest current bid is through the ask price, which is the lowest value that will be accepted for the next bid. Although the bidders cannot determine what the highest bid is (i.e., not strictly speaking an open auction), they do have some information about the current state of the auction (i.e., not strictly speaking a closed auction). As a result, eBay lies somewhere in the middle, with partial feedback given to each bidder throughout the process. For more information on eBay auctions and an example of how they work, check out Q4.1 and Q4.2 on the book's website.

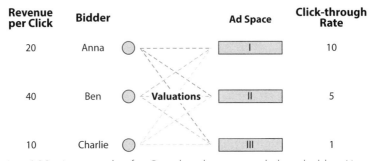

Revenue per Click	Bidder		Ad Space	Click-through Rate
20	Anna		I	10
40	Ben	Valuations	II	5
10	Charlie		III	1

Illustration 4.11 An example of a Google ad auction with three bidders (Anna, Ben, and Charlie) and three ad spaces (I, II, and III).

Generalized Second-Price Auctions

Given that search ad companies like Google have multiple ad spaces to sell, their auctions are actually multi-item auctions with multiple items (i.e., ads) available to each bidder. Let's try a straightforward way to extend what we have seen for single-item, sealed-envelope auctions to the scenario for Google AdWords.

Illustration 4.11 shows the case of three advertisers (i.e., bidders) and three ad spaces. Each of the bidders, Anna, Ben, and Charlie, has a different expected revenue per click, and each of the ad spaces, I, II, and III, has a different click-through rate. The valuation of a particular ad space to a bidder is the click-through rate times the expected revenue per click. There's a total of nine different valuations; for instance, Ben's expected revenue (per hour) for space III is $40 \times 1 = $40.

What does each advertiser provide to Google in order to participate in the auction? Their revenue per click (or more precisely, the amount they value each click). Interestingly, Google will only take this single number from each advertiser, rather than a separate number for each ad space. This would suggest that a click is worth the same to a buyer regardless of which space it came from. But a buyer may, for example, value the first ad space at $100 per click and the second at only $95 per click, because clicks on spaces closer to the top may have a higher chance of leading to a purchase. For simplicity, the sponsor search industry feels that the single number suffices.

The multiple-item auction will allocate a separate ad to each advertiser (if there are more buyers than spaces or vice versa, this just leaves some

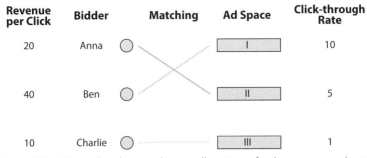

Revenue per Click	Bidder	Matching	Ad Space	Click-through Rate
20	Anna		I	10
40	Ben		II	5
10	Charlie		III	1

Illustration 4.12 Generalized second-price allocation of ad spaces to advertisers for the example in illustration 4.11.

advertisers with no space or some spaces empty, respectively). For ad space allocation and charging, Google AdWords uses a **generalized second-price** (GSP) auction:

- *Bid by advertisers*: Each buyer submits a bid, which is the price they are willing to pay Google per click.
- *Matching by Google*: The ad spaces are allocated to buyers in descending order of their bids. So, the highest bidder gets the first ad space, the second highest gets the second space, and so on.
- *Payment from advertisers to Google*: The price each buyer pays for the space follows the method we saw for second-price, single item auctions. The highest bidder pays the price that the second-highest bidder would pay for the first space, the second-highest bidder pays the price the third-highest bidder would pay for the second space, and so on.

With this method in hand, let's find the outcome of the auction in illustration 4.11. Assuming that each bidder bids his/her true valuation, Anna, Ben, and Charlie will submit $20, $40, and $10 per click, respectively, as their bids. Which spaces will be allocated to which bidders? Under GSP, the highest bidder gets the most-valuable ad space, the second-highest the second-most-valuable, and so on. So, Ben gets the first, Anna the second, and Charlie the third, as you can see in illustration 4.12.

What will each of the buyers pay Google? With GSP, a buyer pays the price the next-highest bidder would have paid for her space:

- Ben is charged $20 per click (Anna's bid). With a click-through rate of 10 per hour, his payment amounts to $20 \times 10 = \$200$ per hour.

Buyer	Revenue	Ad	Valuation	Price	Payoff
Anna	20	II	100	50	50
Ben	40	I	400	200	200
Charlie	10	III	10	3	7

Illustration 4.13 Summary of the generalized second-price ad auction example from the point of view of the buyers. The revenue is per click; the valuation, price charged, and payoff are per hour.

- Anna is charged $10 per click (Charlie's bid), which comes to $10 × 5 = $50 per hour with a click-through rate of 5.
- For Charlie, there is no "next highest" bidder. For this case, Google resorts to its standard minimum bid. If this is $3 per click, then Charlie will pay 3 × $1 = $3 per hour.

How much money will Google make from the auction? Adding the buyer's payments, Google will collect $200 + $50 + $3 = $253 per hour from the auction, as long as the click-through rates turn out to be just as expected. How about the payoffs of the three advertisers? Remember from page 78 that payoff is the difference between valuation and price paid. We already have the payments calculated, so we just need the valuations:

- Ben's valuation of ad space I is the product of his revenue per click ($40) and the click-through rate of this space (10), which comes to $40 × 10 = $400. So, his payoff is $400 − $200 = $200 per hour.
- Similarly, Anna's valuation of the second ad space is $20 × 5 = $100, so her payoff is $100 − $50 = $50 per hour.
- Finally, Charlie's valuation for ad space III is $10 × 1 = $10, so his payoff is $10 − $3 = $7 per hour.

In summary, the total payoff is $200 + $50 + $7 = $257 per hour (again, as long as the click-through rates turn out as expected). Illustration 4.13 summarizes the outcome from the buyers' perspective.

GSP is the method Google uses to determine a ranking of ad space bidders—in turn dictating how they will be ordered on the search results page for the relevant set of keywords—while maximizing profit. However, GSP is not the only mechanism that can be employed to determine a matching. Others, which we don't discuss, may yield a different result. In fact, for multi-item auctions, GSP does not encourage truthful bidding, even though it does for single item auctions. Another method, called **Vickrey-Clarke-Groves** (VCG), will encourage it in either case. But there are disadvantages to that method as well. The question of what the "right" ranking is can have more than one answer.

We should mention that when describing the different types of auctions in this chapter, we have made a few simplifying assumptions about the nature of valuations, revenues per click, and click-through rates. If you're interested in learning more about those, check out Q4.3 on the book's website.

So we now understand how Google generates a large portion of its revenue. Next let's explore how it ranks its standard webpage results to make searching as efficient and high quality as possible.

5

Ordering Search Results

There's no doubt about it: you and almost everyone you know uses Google search. Type a phrase into the search query, hit enter, and out pops perhaps hundreds of millions of relevant links. It's likely that you'll find roughly what you're looking for in the first few results.

The amount of information on the Internet has been growing at a rapid rate since the introduction of the web in 1989. It's hard to estimate exactly how many unique webpages exist today, but current statistics place the number at possibly 60 trillion (that's 60,000,000,000,000!).

How do search engines like Google keep track of all these pages? Each engine has its own database that stores information about all webpages it knows of. With the web growing and evolving as fast as it does, how do they keep their databases up to date? By constantly crawling the web, through programs that automatically follow links from one webpage to the next, adding new pages to the database in the process and updating entries of existing pages. There's no guarantee that this crawling process will reach all the pages out there, though.

The size of Google's index has grown drastically over time. In illustration 5.1, you can see the change from when Google was a prototype at Stanford in 1997 to when it stopped publicizing the number of pages in its database on its home page. The index size has increased exponentially over two decades, from 24 million back in 1997, to 8 billion in 2005, to 60 trillion in 2015.

If Google has indexed trillions of pages, how is it that when you enter a search query, you can usually find more than enough of what you need within the first few results? This is the question we are interested in here. Clearly, the pages aren't displayed in order of when Google first indexed them. In fact, the search engine employs its own famous algorithm—PageRank—which solves a huge system of equations to determine the importance of each page, and then ranks the results relevant to a query from highest to lowest importance.

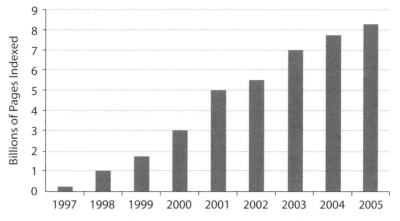

Illustration 5.1   The number of webpages indexed by Google from its origins in 1997 until it stopped posting the number on its home page in 2005.

RELEVANCE VERSUS IMPORTANCE

The idea of ranking webpages dates back to the early 1990s, when the first search engines were created. Naturally, the engines at this time were much less sophisticated than they are today. Due to storage and computational limitations, their databases were quite small and would store only subsets of pages, like the titles and headings of different sections. In this way, searches could be performed quickly and inexpensively.

What was the implication of this abstraction? A major loss of information and search precision. For example, suppose part II of this book was a webpage on the Internet. In building a database entry for it, the first search engines would probably have just stored the titles (e.g., "Ranking Is Hard," "Open Auctions," "Ordering Search Results," "Relevance versus Importance") to represent it. Then, the words in a user's search query would be compared with the words in this database entry to see whether there was a match. So if a user searched for "Auctions," it would find a match to part II. But if the query was "search engine," no match would be found, even though it's a significant part of our discussion here.

Advances in technology soon made **full text search** possible. With full text search, every word of content on a webpage is stored in the database, allowing search queries to be matched against all the content. The first well-known search engine providing this feature was WebCrawler in 1994, which was purchased by AOL a year later.

Ranking by Relevance

If you were designing a search engine, how would you rank the webpages? Perhaps, for a given search query entered by the user, you would order the pages that match the query by the number of times they contain the query. After all, more occurrences should indicate a higher match. You can see a (very small) example of this in illustration 5.2: when the user searches for "Vanilla," each page is checked for the number of times this word appears. Four pages labeled A–D contain this word, and the number of occurrences is one, five, nine, and two times, respectively. The search results page will give links to these four pages in the order C, B, D, A (with some text from each page to give a brief description).

This method of counting occurrences is one way to measure how relevant a page is to a query, or how strongly associated the search is with the page. Ranking based on relevance was the approach that was taken by the early search engines to order their results. In other words, they would display more-relevant pages ahead of the less-relevant ones, in an effort to place the most useful results up front.

Adding Importance to the Mix

Is ranking based on relevance (alone) the best a search engine can do? Not if we consider the widespread success that Google search has had over the past two decades.

Google entered the search engine scene in 1997. At that time, its two founders, Sergey Brin and Larry Page, had come up with a novel approach for ranking webpages. To them, the best way for a search engine to accomplish the daunting task effectively was to take into consideration two different factors for each webpage:

- a **relevance score**, based on the concept of relevance, dictating how relevant the content on the page is to a given search query; and
- an **importance score**, measuring how important the webpage is, irrespective of the content on the page and the text in the search.

Computation of the relevance score did set Google's vision apart at least somewhat. In particular, Google began taking such factors as capitalization, font, and position of content into account, which was not done by the other search engines at the time. But it's the notion of importance that made **PageRank**—Google's ranking algorithm (a clever pun on Larry

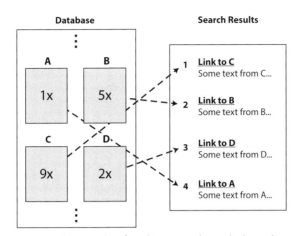

Search for: <u>Vanilla</u>

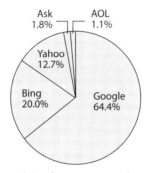

Illustration 5.2 Small example of ranking search results based on relevance.

Illustration 5.3 Market share of the five most popular search engines in the United States in March 2015. Though Bing has been steadily increasing its share over the past several years, Google search still possessed almost two-thirds of the total market and more than three times what Bing did at this time.

Page's name)—much more successful than previous approaches. It has been a driving force in bringing Google to fame since the late 1990s. As of 2015, Google possessed almost two-thirds of the total search engine market, as shown in illustration 5.3.

In the rest of this chapter, we take a look at how importance scores are determined. This involves looking at the network of webpages induced by the hyperlinks that connect pages to one another and form the web in the first place. Keep in mind that unlike the relevance score, the importance of each page will not change depending on the search query entered by the user, nor based on the content that is contained on the page. Instead, it is based entirely on the structure of a graph showing how webpages point to one another.

GRAPHS AND WEBGRAPHS

Webpages are connected to one another through **hyperlinks** (i.e., references to external data in a page that a user can follow). The idea of embedding hyperlinks in text was integral to the creation of the World Wide Web, because hyperlinks are the means by which webpages reference one another. That is, one page points to another if it has a direct hyperlink that will take the user to it.

Connections between webpages can be represented succinctly using a **graph**. (The Internet that enables the web can be described as a graph, too, which we'll look at later.) We could spend an entire book on the mathematics of graph theory, but simple terminology here will suffice: a graph consists of a set of **nodes** (or vertices) that are interconnected by **links** (or edges). In this chapter, we view nodes as webpages and links as an indication of whether one page references another or not. What we are constructing is called a **webgraph**. A webgraph is **directed**, meaning its links are asymmetric: the fact that page A references page B doesn't mean that page B will reference page A.

As an aside, we are going to see many different types of graphs in this book. They differ based on what constitutes a node and what exactly the links between nodes are meant to represent. Beyond the graphs of webpages we look at here, we will consider graphs, for example, of Internet routers (where links are physical connections) in chapter 12 and of human beings (where links are social connections) in chapters 8, 10, and 14.

Webgraphs are essential to understanding importance scores, because they encapsulate the structure of the web's connectivity. Moving forward, we will use the (very small) graph shown in illustration 5.4 to demonstrate the key steps in computing importance scores. In this webgraph, there are four pages (W, X, Y, and Z), and eight hyperlinks. We will assume that all these pages are relevant to the search query the user has entered, so that they will each be displayed on the results page; the question is where.

As another aside, throughout this book, you'll find that we stick to small graphs that can be written on a single page and are easily digestible while

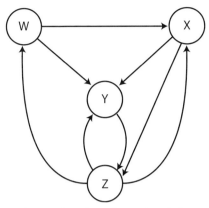

Illustration 5.4   Example of a small-scale webgraph with four pages (nodes) and eight links. The graph is directed, because the links are asymmetric: for example, page W points to page X but not the other way around.

still illustrating the important points. For a moment, though, let's envision what the structure of the entire web looks like. Between these trillions of nodes, we can be sure that the graph is extremely **sparse**, meaning that most webpages are only connected to a tiny fraction of the other pages in the web. Even a large page on Wikipedia with a few hundred links (which is many more than a typical webpage will have) is only connecting to a minuscule fraction of the trillions in total.

What PageRank Does Not Do

So, what makes a page "important"? Perhaps the number of others that point to it? This is called the **in-degree** of the webpage, and measures how many **incoming links** the node has:

- From illustration 5.4, you can see that page Y has incoming links from pages W, X, and Z, giving it an in-degree of 3.
- In contrast, page X has an in-degree of 2, with incoming links from W and Z.
- Page Z also has an in-degree of 2, with links from X and Y.
- Finally, page W has an in-degree of 1, with a link from Z.

By this measure of importance, Google would return Y, X, Z, and then W (with X and Z being interchangeable) as its results, in that order. But does

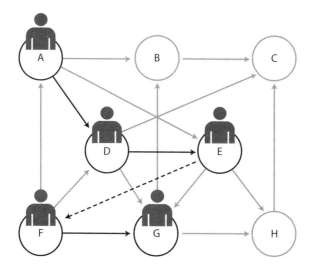

Illustration 5.5 The random surfer philosophy. A person starts on webpage A and randomly clicks on one of the hyperlinks, selecting D. From D she chooses a reference again, ending up at E. Once at E, the person decides to enter a random address into the browser, ending up at F. From F, she clicks on a link to G, and the process continues.

this tell the whole story? Not according to Google's PageRank algorithm. Let's take a look at that now.

THE RANDOM SURFER

Google explains the PageRank concept through an analogy to someone surfing the web at random. This surfer is given a webpage and keeps clicking on links randomly. During this process, she may eventually get bored and enter some other address into the browser. You can see this idea in illustration 5.5: the person goes from A to D and then E based on hyperlinks, then randomly enters F into her browser, and so on. According to PageRank, the percentage of times that a page will be visited in this process (relative to the total number of visits to all pages) is that page's importance.

We'll break down the random surfing process step by step. Let's suppose the user follows the hyperlinks, picking from those available at random. From a given page, the chance that she will select a specific hyperlink, then, gets proportionally smaller as the **out-degree** of the webpage gets larger.

The out-degree is the number of **outgoing links** the node has, as opposed to the in-degree being the number of incoming links.

Back to the graph in illustration 5.4. What is the out-degree of, say, page W? Two. There's a 50% chance a random surfer on page W will navigate to page X next, and a 50% chance she will choose page Y. She can't go directly to page Z, because there's no such link.

Now, what's the chance that the user will be on a given page in the first place? This will surely depend on the in-degree of the page, as we mentioned previously. But it depends on something more: the importance of the links that point to the page. For instance, while Z only has two incoming links, one of those links is from Y, which is the most important in terms of in-degree. If Y is that likely to be visited, surely Z will be at least as likely, because once on Y, the only option the surfer has is to click on page Z!

Quantifying Importance from Graphs

We can represent these concepts of in-degree, out-degree, and page importance visually through the nodes and links of the webgraph. Each node will be given an importance score: we'll call them lowercase w, x, y, and z, for pages W through Z, respectively, as in illustration 5.6. We can equate the importance of a page to the chance that it will be selected in the random surfing process. Each link, with a start node and an end node, can then indicate the chance that a random surfer will click to go to the destination page from the origin page. In other words, a page "spreads" its importance across its outgoing hyperlinks.

For instance, take a look at page X. It could be selected if the surfer is on either page W or page Z. So, we can split this into two parts: the chance that a transition will happen from W to X, and the chance that a transition will occur from Z to X.

First, what is the chance that a transition will happen from W to X? This requires that (i) we transition to X when we are on W, and (ii) we are on W in the first place:

- For (i), we said before that the chance of transitioning to page X starting from W is 50%, or $1/2$.
- For (ii), this is just the importance of W, or w.

Because these events must occur together, we multiply their chances (similar to how we multiplied chances of not transmitting when we looked at WiFi protocols in chapter 2). This gives $w \times 1/2 = w/2$.

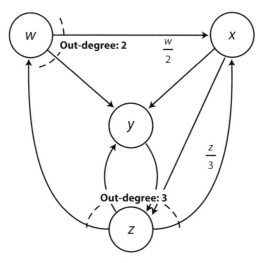

Illustration 5.6 The importance score x for X is dependent on W and Z. Because W and Z have out-degrees of 2 and 3, the weights for these links become $w/2$ and $z/3$.

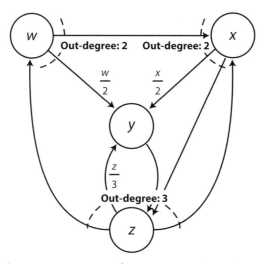

Illustration 5.7 The importance score for page Y is dependent on W, X, and Z. Because W and X both have out-degrees of 2, their links to Y have weights of $w/2$ and $x/2$. Because Z has an out-degree of 3, this link has weight $z/3$.

Second, what is the chance that a transition will happen from Z to X? We can apply the same logic: it requires that (i) we transition to X from Z, and (ii) we are on Z in the first place:

- For (i), X is one of three possibilities, so the chance is 1/3, or 33.33%.
- For (ii), the chance is just z.

Multiplied together, this makes $z \times 1/3 = z/3$.

Since we can arrive at X from W or Z, we add these chances together to get an expression for X's importance score: $x = w/2 + z/3$.

What about for page Y? It can be selected from any of the other pages in the graph, because they all point there. While at W, there's a 50% chance of picking Y; from X, there's also a 50% chance; and from Z, there is a 33.33% chance. So, $y = w/2 + x/2 + z/3$, as shown in illustration 5.7.

Applying this logic to each of the four webpages in illustration 5.4, we come up with the following equations that capture the relationships among the importance scores:

$$w = \frac{z}{3}$$

$$x = \frac{w}{2} + \frac{z}{3}$$

$$y = \frac{w}{2} + \frac{x}{2} + \frac{z}{3}$$

$$z = \frac{x}{2} + y$$

As you can see, a webpage's importance score is dependent on other webpages' scores, and the other webpages' importance scores are in turn dependent on the original's. This type of seemingly circular logic requires us to solve a system of simultaneous equations. The webgraph is an easy way to visualize the equations, by following the procedure that we went through in illustrations 5.6 and 5.7:

1. Label each link as the origin node's importance divided by its out-degree.
2. At each node, set its importance score equal to the sum of all the values on incoming links.

In the end, the number of equations you have will be the number of nodes in the graph.

The Solution

Back to the system of equations above, what we have now are four equations and four unknowns (w, x, y, and z). Is there a set of importance

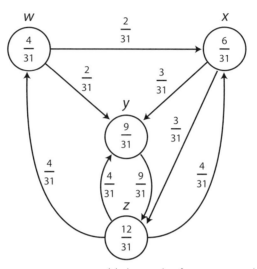

Illustration 5.8 Importance scores and link weights for our example webgraph. You can see that each page's importance is the same as the sum of the incoming as well as the outgoing importance scores of the page.

scores that satisfy each of the equations? In general, there are many ways to solve a system of equations like this. For very large webgraphs with many variables (e.g., the 60 trillion in Google's index), engineers have come up with sophisticated ways of finding the answers quickly and efficiently.

If you are interested in how to solve this particular system by hand, using nothing more than basic algebraic manipulations, check out Q5.1 on the book's website. We can easily verify that the following is the correct answer:

$$w = \frac{4}{31} = 0.129$$

$$x = \frac{6}{31} = 0.194$$

$$y = \frac{9}{31} = 0.290$$

$$z = \frac{12}{31} = 0.387$$

How can we tell for sure? Just plug the values back into the equations to make sure they are all satisfied. For example, the third equation needs $y = w/2 + x/2 + z/3$. Does this hold? Yes: the right side adds up to 2/31 +

Page	In-degree		PageRank	
	Importance	Rank	Importance	Rank
W	0.125	4th	0.129	4th
X	0.250	2nd (or 3rd)	0.194	3rd
Y	0.375	1st	0.290	2nd
Z	0.250	3rd (or 2nd)	0.387	1st

Illustration 5.9 Importance scores and rankings computed for the webgraph in illustration 5.4. PageRank leverages the connectivity in the webgraph, whereas in-degree doesn't.

$3/31 + 4/31 = 9/31$, which is y. You can go through the remaining three equations in the same way.

We can also visualize the solution by filling out the different components on the webgraph. You can see this in illustration 5.8. For each webpage, the following three things are equal: the importance score of the page, the sum of the incoming importance, and the sum of the outgoing importance. For example, at page Z, $12/31 = 9/31 + 3/31 = 4/31 + 4/31 + 4/31$. At page X, $6/31 = 2/31 + 4/31 = 3/31 + 3/31$.

So, what is the ordering of pages based on these importance scores? The ranking from most to least important is Z, Y, X, W. With the naive in-degree method we looked at earlier in the chapter, the ranking was instead Y, X, Z, W. With PageRank, Y is no longer the most important, and Z has become the most important. Even though page Y has the most incoming links, PageRank also takes into account the "magnitude" of these links: in particular, two of Y's incoming are from the less-important nodes W and X, with only half of their importance being spread to Y anyway, and even though the other link is from Z, this only pulls one-third of Z's importance.

Also, page Z may only have two incoming links, but one of these pulls the entirety of Y's importance, because Y points to no other nodes. Because Z also gains importance from X, it is automatically higher up than Y. Referring back to the random surfer concept, this makes sense, because whenever the surfer lands on Y, she will go to Z next, but not the other way around. Over time, Z will get more visits than Y, and therefore should be ranked as more important.

You can see a summary of the importance scores and rankings for each method in illustration 5.9. The in-degree importance here is obtained by dividing each page's in-degree by the sum of the in-degrees across all nodes

(8), simply to normalize the result. So, for pages W, X, Y, and Z with in-degrees of 1, 2, 3, and 2, we get $1/8 = 0.125$, $2/8 = 0.25$, $3/8 = 0.375$, and $2/8 = 0.25$.

Is the Ranking Robust?

What we've looked at is a method that allows us to find a set of importance scores that is consistent across various equations, thereby achieving a ranking. However, are we always guaranteed a unique ranking for any webgraph?

The answer is: not quite yet. In general, we have to make two modifications to our procedure to guarantee a single, unique solution. These involve dealing with cases in which a webgraph has **dangling nodes**, which are nodes that don't point to any other nodes, and in which it has multiple **connected components**. For more information on these special cases and how PageRank accounts for them, you can take a look at Q5.2 and Q5.3 on the book's website. Suffice it to say that these modifications involve accounting for portions of the random surfing process that we didn't describe in detail here. If we apply these fixes to our procedure, then PageRank will always have a unique answer.

The specific procedures that can be used to make the PageRank computation scalable to trillions of webpages require some advanced mathematics that go beyond our scope. But you now have a conceptual understanding of how Google ranks the webpages on its search results page. Though our focus here has been largely on how to get the importance scores, don't forget that the relevance scores are also a significant ingredient to the search engine recipe, so that irrelevant pages (i.e., those without content matching the user's query) can be filtered out in the first place.

Summary of Part II

Here in part II of the book, we have explored methods of ranking, which happen to be fundamental to Google's operation. We have investigated two key cases in which the company sifts through information about particular items—whether of bids or of a webgraph—to find some effective ranking of the items. Illustration 5.10 summarizes the two cases: in auctions, AdWords uses generalized second-price auctioning to determine how to best match bidders to ad spaces, and in PageRank, Google extracts importance (and relevance) scores from enormously large webgraphs to determine which pages to display where in their search results.

Ranking appears in many networking contexts beyond search engines, ad spaces, and Google. It is no wonder, then, that finding a ranking is such a difficult, yet imperative, part of networked life. In chapters 6 and 10, the principle of ranking will come up again, for product lists and graphs of people in social networks, respectively.

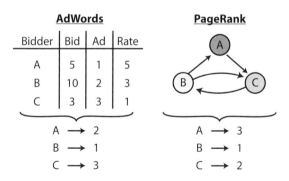

Illustration 5.10 Summary of the two different case studies of ranking that we looked at in this part of the book.

A Conversation with Eric Schmidt

Eric Schmidt is the executive chairman of Alphabet Inc. (formerly Google). From 2001 to 2011, he served as the CEO of Google.

Q: Let me start by asking you this question. Can you imagine a world without Google? What might it look like?

Eric: So I remember a world without Google, and knowledge was largely anecdotal or learned in books. And what happened was like this: because you watched television and you read a newspaper and so forth, you had a general knowledge about things. People would come and tell you things, and you had no way of verifying them. We're verifying them now. People say things to me all the time that may or may not be true, and I check them using Google, and that's indeed how Google itself works. So when you're inside of Google and somebody makes some statement like, did you know that the moon has recently been proved flat, you would then type that in and you would see that in fact that's not the case. So once you develop this lifestyle of when anybody says anything, go ahead and check it, it's a trust-but-verify model of life, which is not a bad way to live.

Q: Do you think this is dependent on the basic assumption of wisdom of the crowds, because Google goes out to get everybody's publicly disclosed information, therefore the chance that everybody is getting it wrong at the same time is very small?

Eric: The observation about Google is that, as a general rule, the algorithms are accurate enough that eventually what most people think is true becomes the top ranking. So using the world-is-flat example, if you type "the world is flat," it is actually true in the sense that that's the title of a book. So if we're talking about in the context of *The World Is Flat* book, well, then, it's a true statement, and otherwise it's a false statement. And I find in general the top ranking of Google tends to be pretty accurate, because most people believe in facts. And the best thing that you can do is when somebody sends you something, like did you know that 90 percent of the people in Congress are criminals, you type that in and the first result that comes out is a Snopes result—Snopes is the site that tries to debunk common myths—and you'll read that it's not in fact true.

Q: Aside from information collection and presentation, can you imagine other parts of our life without Google today?

Eric: We want the things we do to be as common and routine as using your toothbrush. You don't think too much about your toothbrush, you just use your toothbrush, so storing your information in the cloud, using Gmail, using Chrome, that's our aspiration. So I, of course, use all of those tools. I keep everything in the cloud. My favorite current example is photos. Like most people, I have a gazillion photos in many, many different places. And because of Google Photos, I can now upload them into my Gmail account essentially and it organizes them, it deduplicates them, and you can search for things that are in the pictures. It uses machine learning and vision to look over the pictures.

Q: Speaking of machine learning over big data, just how individualized do you think recommendations can become down the road—a recommendation of what to watch in terms of video content, a recommendation of friends, a recommendation of what to do today? How individualized do you envision this is going to be? How far away are we from having machine intelligence that completely understands each of us better than we do?

Eric: I think that there're two different questions there. How close are we to getting very, very good recommendations for people? And the answer is, we're pretty close. The reason is that people tend to follow similar patterns, their friends tend to be similar. We always say that everybody is different—you have different friends, but in fact your friends tend to have the same cultural views, the same language obviously, often the same age, same life experiences, and so your friends tend to be pretty good predictors of the kinds of things that you like. It's not 100%, but it's pretty high. So when we do big data and we look for recommendations, for example, on YouTube, they tend to be pretty good.

Now, it's important to know that the person should opt in for that from a privacy perspective, but assuming that you have chosen to get these recommendations, I find them very useful. It does not mean that [the software] becomes some intelligence that's talking to you. That's a huge leap. I think what we know at the moment on that is we know that computer vision is better than human vision, and so things like the Google Photos example work really well, and I think it's a matter of conjecture how much farther that can go. Everyone would like to have sort of somebody that you could talk to that was a robot that was constantly helping you out. The question is, how soon will that occur? I would say that eventually I believe that it will be possible to have a fairly good working personal assistant: "Eric, you need to get on the phone to call the professor at Princeton and you're late, as usual." That kind of interaction, I think, is probable. I think beyond that, it's more a question of speculation.

Q: Do you think there might be a trend that people are understood to behave in a certain way and are fed with the kinds of things to do, content to read and watch accordingly and so it's more and more about "what it is" and less about "what it might be"? In other words, the machine intelligence is actually reinforcing the existing personality rather than giving them a chance to try something completely different and branch out?

Eric: This question has been asked in many different forms over the last decade, and so far there's no evidence that this sort of bias exists. If what people talk about is causing people to become more narrow, I would argue that life online in the big data world means that you're exposed to a very, very large number of things that you weren't before.

So let's go back to life before Google. Here you are, and you went to a school, you have a job, you have a family, you watch TV. That's not a lot of viewpoints, whereas today when you spend your time on Twitter, Facebook, using email and so forth, you're exposed to a much, much broader set of claims—some of which are false, by the way, and then you use Google to debunk them. So I disagree—I understand the argument, but there's no evidence of it, and there's a lot of evidence that being online at the moment, people are being exposed to a very large number of new things, which is, in many cases, destabilizing. They don't want to hear it or they don't agree with it.

Q: Indeed on YouTube, I do see both similar content repeatedly recommended to me, but also brand-new content that I never thought I would ever see.

Eric: And it's possible for these systems to get pretty good at generating what we call serendipity, things which you didn't know you might be interested in but it can discover, because it knows a little bit about the things you care about. And as you know, technologically all these things are big neural networks that have been trained, and so they're not going to be totally precise. But for general observational knowledge and being generally familiar with things, they'll be perfectly useful.

No one's arguing that these things are going to be right all the time. No one's making the claim that somehow they're always truthful. We don't make a claim that Google tells the truth. We make a claim that Google tries to rank the information as best we can to produce the best answer in the situation, and we have done a good job of that.

Q: Speaking of communication among people, I guess in the good old days there was just the telephone, and then there was something called email, and then that's about it. And then there's texting and then there's social networking and then there's taking a picture, sending it to others, and then there's WeChat, WhatsApp, and so on. It's quite interesting to see that engineers worked very hard to get us to synchronous video chat, and now what people enjoy most is asynchronous, sending text and voice messages to each other. What would be, in your mind, an ideal way to communicate with each other?

Eric: Do you remember the old argument of bowling alone? And the argument was that society would become all the people who were isolated sitting on their sofas watching television and not doing anything social. That was called bowling alone. The only problem with that argument is it's completely false. If you look at society today, people are over-communicating all the time, and I think that that's the lesson from technology, is that people want to communicate all the time in every conceivable way, and that you'll see a multiplicity of ways that they communicate. I don't think there's going to be one way. To argue that, for example, email was the best form of communication is not to understand that there are different forms.

I'll give you an example. WeChat has a very nice texting feature. You can run programs. You can also press a button and leave a short voicemail for the other person. These are very, very rich communication environments, and that's a peer-to-peer one. So I think you're going to see lots of different forms of communication. So while I'm talking to you, I'm getting emails, I'm getting texts, I'm getting voicemails, and there are people outside who want to interrupt me. That's normal in today's world. I'm not suggesting it's good for humans, but it's true. One of the more interesting statistics is that, in a given week, someone touches their phone 1,500 times. The average teenager sends more than 100 texts a day. I just read an article that said that WhatsApp is 50% larger than the total number of SMS messages sent. So the revolution in communication is deep and profound. It's a very, very, very large explosion of communications in all forms. And again, you hear people complaining about this, but we're human beings. All we do is communicate. That's what we've done since caveman days.

Q: Yes. Communication also carries a price now. It used to be you pay for the number of minutes you talk on the phone, and get unlimited data on your mobile device. Now, it is, at least in the U.S., completely switched. You can talk as many minutes on the phone—using the phone as a phone, that is—but then you've got to pay a tiered pricing, maybe a family pricing plan, for your mobile data.

Eric: The original Internet did not charge for things, and whenever you charge for something, you sort of create a scarcity. And the telcos charge for bandwidth, because they have to buy expensive bandwidth from the government and it's hard to lease, hard to operate—so they have a real capacity constraint. There is a new proposal from the government called shared spectrum, where phones would go on and off a shared spectrum, rather than having a single carrier own a large block of that, and that would materially free up bandwidth.

Q: So you're saying that the opportunity for an opportunistic access to spectrum, rather than the static allocation, could optimize the efficiency of using the spectrum to a fuller extent, thereby driving the cost down?

Eric: Right. If you look at the spectrum today, it's essentially empty all the time everywhere. The analogy that we use is that we have highways that are owned by each McDonald's franchise which take you to your McDonald's and nowhere else. It makes no sense. You should have sharing. So sharing makes perfect sense from the standpoint of bandwidth, and it will happen.

Q: With Nest and Glass and other Internet-of-Things [IoT] devices on the rise, people are looking at smart home, smart city, smart factory, and people look around them and they see that buttons can be computers and the cloud might be descending to be right around us like a fog. But people are also worried about privacy and security, especially around the edge of the network with these consumer electronic devices proliferating to be connected. What do you think about security in IoT?

Eric: Well, as a general comment, we should be able to secure this stuff pretty well, and with modern encryption and modern algorithms, your communications should be highly personal, highly secure. And the way this is done is with 2,048-bit encryption, elliptical curve technology, and encrypting everything at rest as well as in transit, so the problems that you see are because people have not done that fully and they need to.

Q: What would you recommend, then, if some of these devices are not powerful enough computationally or energy-wise to run the state-of-the-art encryption? Are there other solutions?

Eric: Again, I disagree with the premise. The algorithms—all of the web now uses HTTPS, which is SSL-based HTTP. There's a great controversy, because the Apple iPhone is so secure that the government can't just seize it and unblock it, and I'm sure you've read about that. The computers are fast enough and the networks are fast enough to accommodate the demands that you're describing without limitation. Of course, there are eventually limitations, but we haven't seen them. My favorite example here is, what is the primary user of the Internet? Video. How in the world could you imagine having television delivered point to point over the Internet? Well, somehow it happens. That's what Netflix and YouTube do, and they work pretty well. It's shocking.

Q: Yes indeed. Speaking of YouTube, that certainly has been and continues to be a very successful deployment, and there are many other examples in what Google does. I read your book, and I'm just wondering if in your mind there is something special about Google's way of innovation.

Eric: Well, if you go back to the core formula, the company is run by technical people to focus on really, really big-bet technologies, and we have a very, very high hiring bar, and we're fortunate that our advertising system produces a lot of profits, which we can then invest in these new things.

Q: You mentioned a high bar of recruiting. Do you think there are enough [talented individuals] in the United States to hire at this point?

Eric: Well, in the first place, the goal of the United States should be to get all the really smart people from other countries and have them move to the United States. Unfortunately, our government has a really stupid policy, which is this H-1 visa limit. So as you know, we educate the best and the brightest, then we kick them out [of the country]. That's really, really stupid. So I think you can never have enough of the kind of people who will invent the future.

Q: I wanted to ask one more question here. Looking at your time at Google, if you had to name a single most critical decision you've made, what would that be?

Eric: It's hard to know, because when you're running a company, you just run as fast as you can, but I would say that the most important thing that we did is we built a system that systematized innovation. In other words, we had lots of ideas, and we kept reviewing them, and we picked the ones that worked. And so you can systematize innovation. You can't predict it but you can systematize it, and you can build things which scale. The combination of the innovation and the scale, just great products that grew quickly, brings you very fast growth. And if you want to look at non-Google examples, take a look at Uber—a relatively simple idea. It's complicated underneath. But once you get it right, it kind of works everywhere, so they can scale very quickly all throughout the world unless they're prevented by regulations or governments. But the product works, and once it works in one place, you can do it pretty much anywhere.

Q: This reminds me of how Henry Ford systematized the assembly line in manufacturing. When you say systematize the process of innovation, how would an innovative idea be evaluated and be either thrown out or advanced or reevaluated later? How does that happen?

Eric: We had a thing called 20% time, where people were encouraged to spend one-fifth of their week on things of their own interest that they were curious about, and many of the ideas started at 20% time. That's the good news. The bad news is that today it takes 100 people to do a product. These products are big and complicated. But they always start with a single or a small-team set of ideas, and they get excited and they run at it. So the more of those starts that you can create with highly technical people who understand where the technology can go, the more innovative you'll be.

Q: How many rounds of scrutiny and refinement does it usually take for that initial idea to progress from a 20% time exploration to a decision that this is going to be a product?

Eric: I don't think that there was a single rule. There were things which went faster and things which went slower. But the key thing to do would be to constantly be reviewing them and seeing if they were making progress. And some things worked, some things didn't. That's fine. And obviously once it becomes clear that it's not going to work, the quicker you can cancel it and recycle the team, the better.

Q: I figure people, once they're used to this process, would actually enjoy that process?

Eric: It's pretty rough. People don't like their projects failing, they don't like the reviews. It's not easy.

Q: Well, so far it's been working just fantastic. Thank you, Eric, for sharing your thoughts.

PART III
CROWDS ARE WISE

Retail shopping, movie watching, and taking classes are just three of our common daily activities that have been impacted by the Internet. Thanks to eCommerce sites like Amazon, content distribution sites like Netflix, and online course sites like MOOC (massive open online course) providers, we can do them from the comfort of our own homes.

By doing these things online, we are contributing to an ever-increasing body of knowledge about peoples' behaviors and preferences. As we navigate through websites, our actions are typically being stored and—in many cases—used to change the experience of those who subsequently visit the sites. For example, when you leave some feedback for a product on Amazon, it may affect where this product appears in some rank-ordered list of items on an Amazon page. When you rate a movie on Netflix, it could affect whether this particular movie will be recommended to someone else.

In this part of the book, we explore the ideas behind how products are ranked on Amazon (chapter 6), how movies are recommended by Netflix (chapter 7), and how people learn from one another in MOOCs (chapter 8). At the innermost workings of these applications is the "wisdom of crowds," which is the notion that as a crowd gets larger (i.e., as more information about a product or item is gathered), the collective decisions made by the crowd will be better (i.e., the estimate of the quality of the product or item will be more accurate). Then, in part IV, we'll turn to the not-so-wise aspects of crowds.

As our journey into social networks begins, we do have to caution you that modeling these types of networks is generally much trickier, with substantial gaps between models and reality. In both parts III and IV, we have to be careful with limitations of the models in both their explanatory and predictive capabilities and their underlying assumptions. And we have to be sensitive to potential abuse of terms such as "averaging" and "probability."

6

Combining Product Ratings

A growing amount of retail shopping, whether it be for shoes, DVDs, text-books, or other commodities, is done on the Internet. In 2014, people spent a combined total of $1.3 trillion buying items online, which was about 6% of the total retail market at the time, promising even more room to grow. By 2018, this spending is projected to nearly double. With so many products and options to choose from, giving customers powerful indications of "quality" to guide their decision making is an essential part of online retail's success.

THE "RIVER" RUNNING ECOMMERCE

The largest eCommerce company in the United States is Amazon (illustration 6.1). Founded by Jeff Bezos in 1994 as an online bookstore, the company pursued a rather unusual business model initially, not expecting to make any profit until the turn of the century. Although its initial growth was slow, Amazon was one of the few eCommerce companies to survive when the dot-com bubble burst around 2000. In fact, the company turned its first profit soon after this, in the fourth quarter of 2001.

Fast-forwarding to today, Amazon is making ever-rising revenue each year, and anything from clothing and shoes to software and electronic devices can be purchased from its site. You may have even purchased this book from there!

Amazon's services have also branched outside eCommerce. In 2007, the company moved into production with their own electronic book (e-book) reader, the Kindle (which we mentioned in our discussion of smart data pricing in chapter 3). Three years later, they announced that Kindle sales had outgrown sales for hardcover books on their site.

For many years, there was a clear distinction between the street-side, brick-and-mortar retailers that deal with customers face to face and online

Illustration 6.1   Amazon's trademarked logo.

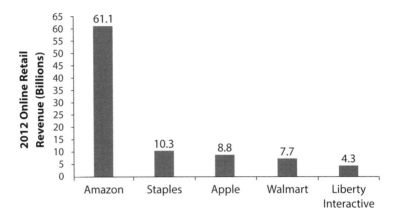

Illustration 6.2   Total revenue generated from online sales by the largest five online retailers in 2012.

retailers like Amazon. Then, many in-store retailers expanded to take advantage of the burgeoning eCommerce industry. A prime example is Walmart, which is the largest brick-and-mortar retailer and is consistently among the top five online. In 2012, Amazon's revenue from online sales was still more than eight times as large as Walmart's, as you can see in illustration 6.2. In the second quarter of this same year, it was estimated that Amazon received 100 million unique website visitors, more than doubling that of Walmart.

So, why has Amazon been so successful? Aside from its many features (e.g., large selection, free two-day shipping in some cases) and competitive prices, part of what makes the site so attractive is that it has a mechanism for customers to provide feedback: through product reviews. Amazon will in turn aggregate the reviews on a given product together to form a single number—the average star rating—that gives shoppers a sense of the product's "quality." In this chapter, we explore methods that online retailers like Amazon can use to aggregate opinions, and in turn, to order the list of products shown to shoppers.

Toshiba 32C120U 32-Inch 720p 60Hz LCD HDTV (Black)
Buy new: $379.99 **$249.99**
47 new from $249.99
14 used from $201.95
Get it by Monday, Oct 8 if you order in the next 19 hours and choose one-day shipping.
⭐⭐⭐⭐☆ ☑ (95)
Eligible for FREE Super Saver Shipping.
Product Description – "Toshiba 32-Inch 720p 60hz LCD HDTV"

Panasonic VIERA TC-L32C5 32-Inch 720p 60Hz LCD TV
Buy new: $299.99 Click for product details
15 used & new from $219.95
Get it by Monday, Oct 8 if you order in the next 19 hours and choose one-day shipping.
⭐⭐⭐⭐⭐ ☑ (8)
Eligible for FREE Super Saver Shipping.
Product Description – "…C5 (32-inch class) is a 720p, LCD HDTV (cold cathode backlighting) with …"

Illustration 6.3 Two HDTV listings on Amazon. The one on the right has a higher average rating of 4.5 stars, compared to 4 on the left. But the one on the left has a higher review population of 95 people, compared to 8.

CAN YOU TRUST THE AVERAGE RATING?

Suppose you are shopping on Amazon for a new HDTV. After browsing through the list of search results, you have finally narrowed your search down to two, on the basis of price: the ones shown in illustration 6.3.

Which to choose? At first glance, the one on the right seems like the better choice, because it has a higher rating: 4.5 stars as opposed to 4 stars.

Are we missing something? Let's think about how these numbers were determined. Amazon allows its customers to enter reviews for products they have bought. Each of these reviews consists of three fields:

1. a rating, which is the number of stars given (i.e., 1, 2, 3, 4, or 5);
2. a textual review, which explains why the specific rating was given; and
3. an indication of how many people found this review "helpful."

The so-called average customer review of an item—which is what the 4 and 4.5 in illustration 6.3 are—is the average of all the ratings entered by customers on the item. It is an attempt to summarize the opinions of the reviewers and takes the form of a single number. It's helpful for most people who don't have time to read through each of the individual reviews. As we consider this summary rating, though, shouldn't we be cautious of the people who contributed to it, in terms of their individual reputations and the total number of reviews that were entered?

Dissecting the Average

In illustration 6.3, 8 people have rated the HDTV on the right, whereas 95 have rated the one on the left. What does this mean for the average customer review? Intuitively, if more people have entered reviews for an item, then the average of their ratings is more trustworthy: it's less susceptible to reviewers who have a tendency toward being harsh (i.e., entering lower-scoring reviews than others) or lenient (i.e., entering higher-scoring reviews) and is less prone to random reviewer spam entered by someone who has not even used the product before. To see this, think about a product that has received one 3-star review. If someone comes along and decides to randomly enter a 5, the average rating will change to

$$\frac{3+5}{2} = 4$$

which is an entire star's difference. If instead we started with 101 3-star reviews, then if someone randomly entered a 5, the average would only change to

$$\frac{101 \times 3 + 5}{102} = 3.02$$

Rounding to one decimal place (as is done on Amazon), the 3-star average will appear exactly the same.

Obviously, not all 5-star reviews are too lenient, and not all 1-star reviews are too harsh. These types of reviews can actually be quite useful to shoppers. In fact, Amazon visually highlights the "most helpful" favorable and "most helpful" critical reviews on its products, thereby contrasting the extreme reviews that people found most helpful overall. You can see an example in illustration 6.4: the two reviews have high variability between their number of stars (4 versus 1), but both are well received given the number of people who found them helpful.

Still, we might suspect that the more reviews there are for a product, the more likely it is that the average of the ratings can be trusted. So looking again at illustration 6.3, we would not be so quick to say that the 4.5-star average customer review based on 8 ratings is better than the 4-star average based on 95.

Review Reliability

Reviews can be untrustworthy. Yet they are important in so many contexts of our lives: from online purchases of every kind, to recommendation letters

The most helpful favorable review	The most helpful critical review
104 of 106 people found the following review helpful	502 of 533 people found the following review helpful
★★★★★ **A good but not great set**	★☆☆☆☆ **They stripped a lot of features out for 2012 over the same 2011 model**
LG's lower-mid level sets have earned a reputation for having low gaming lag, great color accuracy, and the most extensive features and picture options of any sets at or even above their price level. The CS560 series still delivers in these regards, but to a lesser extent than earlier models.	They stripped a lot of features out for 2012 over the same 2011/2010 line/models. The reasons are unclear. I assume to just make more money and fleece their customers even more. They removed so many great features that really made this a great set but they kept the price the same, which I would like to point out is already high.
The styling and build quality of the set are very good overall.	

Illustration 6.4 Amazon highlights the most helpful favorable and most helpful critical reviews for its products.

written by past employers, to course evaluations written by students. What approaches are needed to enhance their reliability?

For one, we need methods that will screen out the "bad" kinds of ratings. Prohibiting people from reviewing anonymously and limiting everyone to one review per item are good places to start. What would happen if Amazon didn't have these mechanisms in place? If Bob were selling a product on Amazon, he could enter many positive reviews on his item to boost his average customer review, signing anonymously each time, so that nobody would know he was inflating his own score. Additionally, if he found a product that was competing with his own for sales, he could enter a slew of poor reviews for that item. Even after these two screening methods, there are many other sources of low quality ratings to consider. For example, spammers may enter random reviews that have nothing to do with the product (e.g., containing links to their own sites).

So before anything else, we need to check the mechanism used to enter the reviews. Who can review? How strongly are customers encouraged to review? Do you need to buy the product before you can review it? Can a person who has just created her account enter a review? Also, what is the range of numbers that can be entered for the review? It has been observed, for example, that scales of 1–10, 1–3, and 1–5 elicit different psychological responses.

These points raise tough questions that don't have unique, "correct" answers. They depend on the type of product being reviewed: movies (e.g., on the Internet Movie Database) are very subjective, whereas electronics (e.g., on Amazon) are much less so, while hotels (e.g., on TripAdvisor) and restaurants (e.g., on Yelp) are somewhere in between. They also depend on the quality of the reviewer: for instance, Amazon rewards people for submitting useful reviews by keeping a "top reviewer ranking," and if a reviewer is highly ranked throughout a year, they will be elevated to the

Illustration 6.5 In 1906, an ox was put on display at a farm in the UK, and roughly 800 villagers attempted to guess its weight. While they were each significantly off individually, the average of their guesses was only 1 pound away from the right answer.

"Hall of Fame." Those in this category should probably be trusted more than those who are not. Still, reputation is difficult to quantify.

With these challenges, you may be thinking that opinion aggregation is unlikely to work well. But there have been notable exceptions.

TWO HEADS ARE BETTER THAN ONE

Back in 1906, an intriguing contest occurred on a farm in Plymouth, UK. At a livestock fair, an ox was placed on display, and the villagers were tasked with guessing the weight of the animal. Each of the 787 participants got a good look at the ox and then wrote down their guess on a piece of paper without talking to anyone (illustration 6.5).

Sir Francis Galton, a famous statistician around this time, ran statistical tests on the results. At first glance, he saw a set of numbers all over the place—from very low to very high—in which not one guess was the true weight of 1,198 pounds. But remarkably, when he averaged the guesses, the result was 1,197 pounds, which is off by less than one-tenth of a percent. Even the median (the middle number of the set) was 1,207 pounds, which is still off by less than four-fifths of a percent.

How could it be that everyone's guess was so far off and yet the simple average was so close to the truth? Several key factors were in play here that made averaging work so well. For one, the task was relatively *easy*: guessing the weight of an ox has an objective answer with a clear numerical meaning. Also, the estimates were *unbiased*: everyone got a good look at the ox, so there was no systematic tendency to either guess too low or too high. Further, the estimates were *independent*: none of the villagers saw anyone else's guess, so nobody was influenced by anyone else (similar to how sealed-envelope auctions work, from chapter 4). Finally, a good number of people participated.

These factors are at the core of Galton's remarkable result, but each is present only to varying extents in review creation.

The Wisdom of Crowds

Returning now to Amazon, our hope is that when we average customer ratings for a product, the result will get close to the right rating. But can we even say that a "right" rating exists? Wouldn't this depend at least somewhat on the specific customer (e.g., a certain line of T-shirts may be attractive to one person, but not to another)? In general, three factors are important to consider when aggregating individual opinions, as we saw in Galton's experiment:

- *Task definition*: A well-defined task with a clear and consistent objective (e.g., guessing a number) is more amenable to opinion aggregation.
- *Independent and unbiased opinions*: Success in opinion aggregation stems not from having many smart individuals who are likely to guess correctly, but from the independence of each unbiased individual's view from the rest.
- *Population size*: Galton's experiment would not have worked as well if there had been fewer people participating.

Is reviewing a product on Amazon well-defined? Not exactly. What precisely constitutes another star in a rating will differ from person to person. Are Amazon reviews independent of one another? Kind of. Even though you can see the existing reviews before entering your own, usually your rating will not be affected by them *too* much. But sometimes reviews are made on an item as reactions to recent reviews—for example, to counterargue or reinforce those points (this is an example of sequential decision making, which we talk more about in part IV). In general, the less well-defined the task is and the more dependent the reviews are, the more "guesses" we need to obtain a trustworthy average. If we have mechanisms in place to detect inconsistent or low-quality reviews, we can reduce the size of the population needed.

When these three factors hold, opinion aggregation works remarkably well. For example, let's say we have 1,000 people playing a "guessing" game

Illustration 6.6 Under the wisdom of crowds, the aggregate knowledge of the many will outperform the ability of the individual or the few.

with a clear, well-defined task at hand. At the end, we collect each of the guesses and average them. As it turns out, we can expect mathematically that the error in this average will be reduced by a factor of 1,000 from the anticipated error in each separate guess:

$$\text{Expected error in the average} \approx \frac{\text{Expected error in each guess}}{\text{Number of people}}$$

provided that all the guesses are independent.

This equation puts into math what we have been discussing so far: the **wisdom of crowds**. As long as everyone guesses independently and without any systematic bias, we can expect that their "collective guess" as a group will improve in proportion to the number of people in the crowd (illustration 6.6). So if there are five people, the improvement will be fivefold, if there are ten people, it will be tenfold, and so on. The crowd is wise, even if no individual person in the group is that wise. In illustration 6.7, you can see an example of applying this principle to a group of guesses coming from five people.

At this point, you may have two questions about this equation. First, why does it have the ≈, rather than an equals sign? This means that the relationship holds approximately, not exactly, and is partially due to the fact that we are dealing with probabilities (for more information, check out Q6.1 on the book's website). Second, what exactly does it mean when we say "error"? The word "error" in this equation technically refers not to the difference between the guess and the actual value, but to the square of that difference. We'll deal with squared errors more in chapter 7.

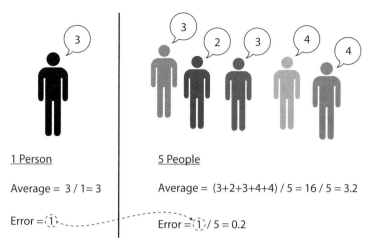

1 Person

Average = 3 / 1 = 3

Error = 1

5 People

Average = (3+2+3+4+4) / 5 = 16 / 5 = 3.2

Error = 1 / 5 = 0.2

Illustration 6.7 We start with a single person on the left, whose expected error is 1. When we move to five people, the error in the average of 3.2 is expected to reduce by a factor of 5, provided everyone's guess is independent and unbiased.

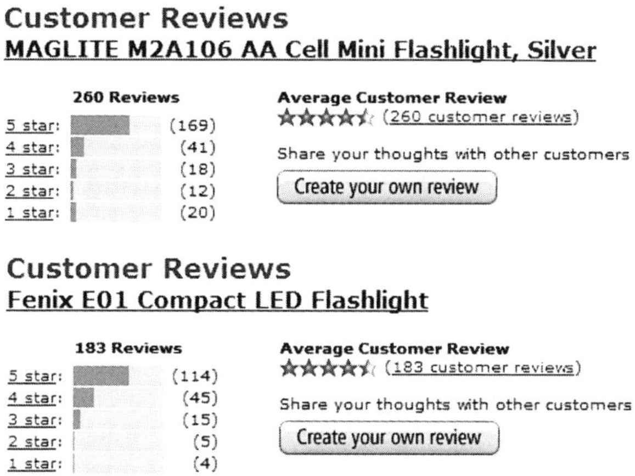

Customer Reviews
MAGLITE M2A106 AA Cell Mini Flashlight, Silver

260 Reviews

5 star: (169)
4 star: (41)
3 star: (18)
2 star: (12)
1 star: (20)

Average Customer Review
★★★★☆ (260 customer reviews)

Share your thoughts with other customers

[Create your own review]

Customer Reviews
Fenix E01 Compact LED Flashlight

183 Reviews

5 star: (114)
4 star: (45)
3 star: (15)
2 star: (5)
1 star: (4)

Average Customer Review
★★★★☆ (183 customer reviews)

Share your thoughts with other customers

[Create your own review]

Illustration 6.8 Two different flashlights offered on Amazon, with similar average ratings and review populations but different review variations.

Rating Aggregation Is Hard

If there are a large number of reviewers for a product on Amazon, does the previous discussion imply that the average rating will be close to the "truth" we want? Not necessarily. We just pointed out a few complications, like the fact that opinions are not (completely) independent.

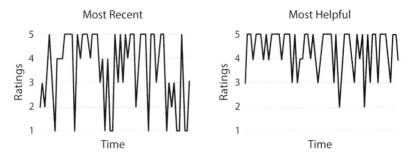

Illustration 6.9 Comparison between the most recent (left) and the most helpful (right) ratings for a product on Amazon.

Other difficulties also arise. Back in illustration 6.3, for example, we have two different HDTVs of comparable price with different average ratings: the Toshiba has 4 stars from 95 ratings, compared to the Panasonic's 4.5 stars from 8 ratings. The customer is faced with a trade-off between choosing a product with a lower average rating and a larger review population versus one with a higher average rating and a smaller number of reviews.

Even if two products have similar average ratings and review population sizes, they may differ in how their ratings are distributed. For example, take a look at the two flashlights in illustration 6.8. Both have an average customer review of 4.5 stars and a similar number of reviews. Whereas 65% of reviewers gave the MAGLITE 5 stars and 8% gave it 1 star, 62% of reviewers gave the Fenix 5 stars and 2% gave it 1 star. So, the MAGLITE has a higher percentage of both high and low ratings. Does a larger variation make the average rating more trustworthy, or less? This is a subjective question with no clear answer.

Finally, the ratings a product receives may fluctuate over time. For example, in illustration 6.9, you can see the 60 most recent ratings compared to the top 60 most helpful ratings over time for a product on Amazon. The average of the most recent ratings is 3.6, substantially lower than that of the most helpful (which is 4.4). Do the most recent ratings reflect a real change—like a new defect in the product—or just normal fluctuations? Which is a better fit for the average customer review? Again, the answers to these questions are subjective, posing another challenge to aggregating opinions.

FINDING A "GOOD" RANKING METHOD

In part II, we looked at different methods used by Google to rank lists of items. Amazon and other eCommerce companies also have schemes in place for ranking lists—namely, lists of products in given categories—to help customers navigate through large sets of items efficiently. We'll introduce one of these methods here, called Bayesian ranking, which has been adopted widely in practice. Amazon does not, however, reveal the details of its own algorithm.

Naive Averaging

Illustration 6.10 shows the ratings for a set of five DVD players on Amazon. The number of stars in each category and the total number of reviews are given. Let's use these numbers first to calculate the average ratings that Amazon will show for each product.

How do we do this? We add together the total number of stars and divide by the total number of reviews. For the Panasonic, there are five 5-star ratings, which is $5 \times 5 = 25$ stars in all; three 4-star ratings, which is $4 \times 3 = 12$; three 3-star ratings, or $3 \times 3 = 9$; and no 2- or 1-star ratings. The total number of stars is $25 + 12 + 9 = 46$, and the total number of reviews is $5 + 3 + 3 = 11$, so the average is $46/11 = 4.182$.

What about for the other four items? You can follow the same procedure, arriving at the following (abbreviating the items by the first letters of their names):

$$S: \frac{18 \times 5 + 9 \times 4 + 5 \times 3 + 2 \times 2 + 3 \times 1}{18 + 9 + 5 + 2 + 3} = \frac{148}{37} = 4$$

$$P: \frac{23 \times 5 + 15 \times 4 + 11 \times 3 + 5 \times 2 + 13 \times 1}{23 + 15 + 11 + 5 + 13} = \frac{231}{67} = 3.448$$

$$C: \frac{18 \times 5 + 14 \times 4 + 2 \times 3 + 6 \times 2 + 11 \times 1}{18 + 14 + 2 + 6 + 11} = \frac{175}{51} = 3.431$$

$$T: \frac{19 \times 5 + 10 \times 4 + 10 \times 3 + 4 \times 2 + 11 \times 1}{19 + 10 + 10 + 4 + 11} = \frac{184}{54} = 3.407$$

If we rank these from highest to lowest average rating, the order is as it appears from top to bottom in illustration 6.10.

Is this the "correct" ordering? Back to our driving question: should a product with only 2 reviews, even though both are 5 stars, be placed higher

DVD Player	5 stars	4 stars	3 stars	2 stars	1 star	Total
Panasonic	5	3	3	0	0	11
Sony	18	9	5	2	3	37
Philips	23	15	11	5	13	67
Curtis	18	14	2	6	11	51
Toshiba	19	10	10	4	11	54
Total	83	51	31	17	38	220

Illustration 6.10 Ratings for a set of five DVD players on Amazon.

than a competing product with 100 reviews that averages 4.5 stars? Intuitively, this would be wrong, and we are facing this problem with the items here: the Panasonic, which has the highest average rating of 4.182, also has the smallest number of reviews.

Bayesian Philosophy for Ranking

How can we account for varying numbers of reviews? We should somehow weight the raw rating scores with the population sizes. Knowing how many reviews there are gives us prior knowledge, and we can take advantage of that.

So in addition to considering each product separately, let's incorporate the information we have about all the products in question. Similar to how we computed the average rating for each, we can find the overall average across the full population of items. What would this be? If we add down the columns in illustration 6.10, we get the total for each star rating, as indicated in the last row. We can then calculate:

$$\frac{83 \times 5 + 51 \times 4 + 31 \times 3 + 17 \times 2 + 38 \times 1}{83 + 51 + 31 + 17 + 38} = \frac{784}{220} = 3.564$$

This overall average is based on the 220 total ratings. Intuitively, we can use it as a backup for each of the individual products. The more ratings there are for a product, the more trustworthy its individual average will be relative to the overall average, in which case we should place more emphasis on the individual. On the contrary, the fewer ratings there are for a product, the less trustworthy its average will be, in which case we should lean toward the overall average rating.

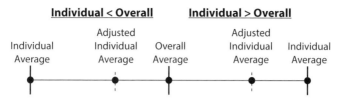

Illustration 6.11   The adjusted average rating sits somewhere between the individual and overall averages. On the left is the case when a product's individual average is less than the overall, and on the right, when it is greater.

We can think of this as a "sliding ruler" between the individual and the total, as in illustration 6.11: the adjusted individual rating lies somewhere between the two. For each product, adjusted value can be determined through this formula:

$$\frac{\text{Overall num.} \times \text{Overall avg.} + \text{Individual num.} \times \text{Individual avg.}}{\text{Overall num.} + \text{Individual num.}}$$

Known as **Bayesian ranking**, this is one of the inference methods under the umbrella of Bayesian statistics. This class of statistics is named after the English reverend and mathematician Thomas Bayes, who discovered a special case of its fundamental theorem in the mid-1700s (most of the heavy lifting in formalizing it was actually done by Pierre-Simon Laplace, who rediscovered it independently and expanded it in the late eighteenth century).

Interestingly, throughout the 1800s and much of the 1900s, Bayesian thinking was largely rejected and even actively suppressed by so-called frequentist statisticians, who strongly preferred the classical approaches to making inferences and estimations from data. But through the years, Bayesian reasoning was used to solve some significant problems that were insurmountable by frequentist methods. Here's a few historical examples:

- When French officer Alfred Dreyfus was falsely accused of treason in the 1890s, mathematician Henri Poincaré invoked Bayes's Theorem to demonstrate his innocence.
- During World War II, British computer pioneer Alan Turing used a Bayesian system to decode the German Enigma military communication cipher.
- In the 1950s–60s, Harvard and Chicago researchers used Bayesian analysis to show that the disputed Federalist Papers were written with high probability by James Madison, not Alexander Hamilton.

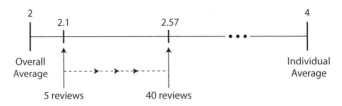

Illustration 6.12 Moving from 5 to 40 reviews in our example causes the adjusted average to move from 2.1 to 2.57, closer to the individual average.

The coupling of successes like these with advances in computing gradually increased the acceptance of Bayesian modeling over time. By the twenty-first century, it had become widely embraced. Today, it is used in fields like machine learning and big data analytics. Let's be clear: Bayesian modeling does carry its own risks, but it sheds some light on the wisdom of crowds that's worth discussing.

Let's say we have 100 total ratings, with an overall average of 2 stars. If one of the products has a 4-star average based on 5 reviews, then what would the Bayesian rating for this item be? Using the equation, we get

$$\frac{100 \times 2 + 5 \times 4}{100 + 5} = \frac{220}{105} = 2.10$$

which is much closer to 2 than it is to 4. This makes sense, because the individual number of ratings is much smaller than the total, so we rely on the latter more. In contrast, if the 4-star average was based on 40 reviews, the Bayesian rating would rise to

$$\frac{100 \times 2 + 40 \times 4}{100 + 40} = \frac{360}{140} = 2.57$$

which is moving away from 2 (illustration 6.12).

Let's apply Bayesian adjustment to our example in illustration 6.10 to see whether the ordering changes. Remember there are 220 total ratings, and we computed the overall average rating to be 3.564. So, what is the adjusted value for the Panasonic player? With 11 ratings for this item and an individual average of 4.182, the Bayesian formula gives

$$\frac{220 \times 3.564 + 11 \times 4.182}{220 + 11} = \frac{784.08 + 46.002}{220 + 11} = 3.593$$

Notice how much closer the Panasonic rating is to the average, 3.564, after Bayesian adjustment. Why? Because of the small number of ratings (11) for

DVD Player	No. Ratings	Naive Average		Bayesian Adjusted	
		Ranking	Rating	Ranking	Rating
Panasonic	11	1	4.182	2	3.593
Sony	37	2	4	1	3.626
Philips	67	3	3.448	4	3.537
Curtis	51	4	3.431	3	3.539
Toshiba	54	5	3.407	5	3.533

Illustration 6.13 Comparison between the rankings of the DVD players based on raw average and Bayesian adjustment.

this specific player relative to the total (220), making the formula weight the overall average much more heavily. The calculations for the other DVD players are done similarly:

$$S: \quad \frac{220 \times 3.564 + 37 \times 4.000}{220 + 37} = \frac{784.08 + 148}{220 + 37} = 3.626$$

$$P: \quad \frac{220 \times 3.564 + 67 \times 3.448}{220 + 67} = \frac{784.08 + 231.016}{220 + 67} = 3.537$$

$$C: \quad \frac{220 \times 3.564 + 51 \times 3.431}{220 + 51} = \frac{784.08 + 174.981}{220 + 51} = 3.539$$

$$T: \quad \frac{220 \times 3.564 + 54 \times 3.407}{220 + 54} = \frac{784.08 + 183.978}{220 + 54} = 3.533$$

What is the result? Now, the ranked order of the products, from highest Bayesian rating to lowest, is Sony, Panasonic, Curtis, Philips, and Toshiba. You can see the pre- and post-Bayesian rankings side-by-side in illustration 6.13. The Panasonic and Sony players, as well as the Philips and Curtis players, have swapped places under Bayesian adjustment. Also, all ratings have been pulled closer to the average.

In this example, Bayesian adjustment changed the ordering of the items significantly. The adjusted ratings will not alter the order in every situation, though. Can you think of an example for which the ranking stays the same?

Bayesian Ranking in Practice

Quite a few websites have adopted Bayesian ranking in practice. The Internet Movie Database's top 250 list of movies, for example, follows the equation four pages earlier exactly. Sometimes, it's better to put some maximum

Customer Reviews
Sony DVP-SR200P/B DVD Player, Black

529 Reviews			Average Customer Review
5 star:	▮	(281)	★★★★☆ (529 customer reviews)
4 star:	▮	(118)	
3 star:	▮	(39)	Share your thoughts with other customers
2 star:	❘	(18)	
1 star:	▮	(73)	Create your own review

Illustration 6.14   An example of a product on Amazon that created bipolar responses: some people liked it a lot, giving it 4 or 5 stars, whereas others hated it, giving it 1 star.

on the value of "Overall num." that is used to adjust the ratings. Over time, the total number of ratings entered across the products is going to keep rising, and the range of values that the Bayesian-adjusted ratings can take will get smaller; eventually, we could reach the point where we have so many ratings for so many products that we are essentially defaulting to the overall average each time we use the equation. Beer Advocate's beer ranking, at http://beeradvocate.com/lists/popular, tries to avoid this by choosing "Overall num." to be the minimum number of reviews needed for a beer to be listed on the page in the first place.

In this discussion we've assumed that there is a single, true value of a product's rating, which corresponds to how desirable we expect it to be to a customer. But in reality, the truth can vary from person to person. Certain products simply create bipolar reactions: some love it, while others hate it. You can see an example of this in illustration 6.14 for a Sony DVD player on Amazon: most people gave it either 4- or 5-star ratings on one hand, or 1 star on the other hand, whereas 2- and 3-star ratings are less frequent. Sometimes, the only people who care enough to write reviews are those who have strong feelings in one of these two directions. Such a collection of ratings is said to follow a multimodal distribution (i.e., one with many centers). Though we have focused on unimodal distributions here, Bayesian analysis can be extended to the multimodal case too.

Finally, we should caution that Bayesian adjustment applies only to adjusting ratings within a comparable family of products (e.g., a set of DVD players, a set of laptops). It cannot be readily used to adjust ratings of products that are likely to be different (e.g., a set of random electronic items). The family of objects being rated as a whole needs to have an overall average that is meaningful for the adjustment to make sense.

At this point, you're probably wondering: what specifically does Amazon do to rank its product lists? They actually follow some secret formula that combines the average reviews with at least three additional elements: the Bayesian adjustment by review population (as we've discussed here), the recency of the reviews, and the reputation score of the reviewer. The exact formula is not known outside Amazon, but if you're interested in walking through a long example of how their rankings may be determined, check out Q6.2 on the book's website.

So far, we've seen how rating aggregation across the reviews that have been entered on a website can be useful when attempting to find the "truth" about products. The wisdom of crowds is an important principle in networking that has guided the design of many algorithms for combining peoples' opinions. Next we turn to Netflix movie recommendations, where—rather than extracting one rating for everyone—we predict several ratings for each person.

7

Recommending Movies to Watch

Continuing with our theme of opinion aggregation, we now turn to *recommendation*. Product recommendation, broadly speaking, uses the existing knowledge about product ratings to advise customers on what to consume next.

If you're a Netflix subscriber (illustration 7.1), you've probably been at the receiving end of the company's movie recommendations. How do they determine which ones (they think) you will be most interested in watching? Making recommendations requires them to make predictions of movie ratings they don't have by leveraging the data they do have about user preferences. This often assumes and leverages the wisdom of crowds: the more ratings they have to start with, the better they can expect their predictions will be.

STREAMING MOVIES AT SCALE

Back in 1997, a man grew frustrated by the large sum of late fees he had racked up from a video store. What he thought was "past due" was a new payment model: rather than charging people per rental, why not charge them a flat rate per month? Other mechanisms, like a cap on the number of movies that could be out at one time, could instead be used to incentivize people to return the DVDs.

Reed Hastings, the man in this story, became one of the founders of Netflix that same year. Netflix initially operated as a traditional brick-and-mortar DVD rental store with a pay-per-rental pricing model. With eCommerce beginning to take off (as mentioned in chapter 6), they opened an online store in 1998, allowing people to shop for DVDs online and wait for them to arrive by mail. Then, in 1999, they introduced their monthly subscription model, where customers would pay Netflix a monthly fee rather than per rental. Under this scheme, a subscriber can keep a rented movie out for as long as she wants, but there is a limit on the number that she can have out simultaneously.

Illustration 7.1   Netflix's trademarked logo.

Over the years, Netflix has been able to operate with great scalability and stickiness. Scalability means that Netflix's cost of onboarding a new customer is much lower when they already have a lot of customers. Stickiness means that those already using Netflix services tend to stay rather than switch to another video provider.

By 2008, there were about 9 million Netflix users in the United States and Canada, a tenfold increase over the number of subscribers it had at the turn of the century. Starting around this time, Netflix migrated toward another form of delivering entertainment: **streaming** movies and TV programs through the Internet to Internet-connected devices. These devices include all those we use every day: TVs, set-top boxes, smartphones, game consoles, and more. This form of rental service has largely eclipsed traditional video stores; for example, Blockbuster filed for bankruptcy in 2010 and has moved on to mail delivery.

Video streaming caused Netflix's subscriber base to skyrocket, increasing more than sevenfold from what it was in 2008 to the 66 million it had become by 2015. Currently, Netflix is the leading video streaming service provider in the United States, with about 36% of all households subscribed to such a service choosing Netflix, compared with Amazon Prime and Hulu Plus (Netflix's closest competitors) at 13% and 6.5%, respectively. Interestingly, Netflix video streaming generated so much traffic in March 2011 that one in every four bits streamed on the Internet was their own. Some statistics have put this fraction even higher.

RECOMMENDATION: A "MIND READING" GAME

Netflix offers a number of features to help you browse through movies. You can filter them by genre, sort them by rating, look at Critic's Picks, consult the Netflix Top 100, and so forth, before choosing.

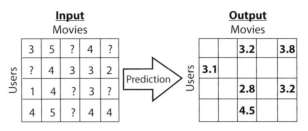

Illustration 7.2 The known ratings of users for movies are input to the predictor, which outputs a prediction of all the unknown values.

Some of these features are "one-size-fits-all," appearing the same for you as they do for anyone else. But Netflix will also recommend movies to you that it thinks you, personally, would like. It's as if they were reading your mind, predicting which ones will fit your tastes before you have watched or rated them!

How does this work? As you browse through Netflix, they build up a history of your actions and behaviors in their database. This information is passed through algorithms to make predictions about how you would rate other movies. As you submit ratings, add movies to your queues, or tell Netflix which ones you're not interested in, these actions are all impacting which ones will be recommended to you in the future.

An effective **recommendation system** is important to Netflix, because it enhances user experience, increases customer loyalty, and helps with inventory control. These systems are also important in many other applications besides video distribution. For example, in chapter 6, we talked about how Amazon comes up with its average ratings and rankings for specific products; a different variation is how Amazon recommends products to you on the basis of your purchase and viewing history, adjusting it each time you browse. Also, in chapter 9, we talk briefly about how YouTube makes recommendations of videos to you at the end of each one you watch.

Netflix has championed a system that goes beyond typical recommendation schemes, using the rich history of all user behaviors it has collected to build a profile of each user's taste in movies and how each movie is rated by users. Its algorithms find patterns hidden in the data provided by the "crowd" of movie raters to build these models. Let's take a closer look at the inputs and outputs to this system, as in illustration 7.2.

The Inputs

Each time a user rates a movie, her rating is stored in the Netflix database. The collection of this data forms the input to the system. Each rating contains four numbers of interest: the user's ID, the movie's ID, the number of stars (from 1 to 5), and the date of the rating.

How big is this input? Very big. Remember that Netflix has more than 60 million users: with about 75,000 different movie titles, there are more than 4 trillion (that's 4,000,000,000,000) possible pairs of users and movies. Of course, only a tiny portion of these ratings are actually known, since only a small fraction of users will have watched a given movie, and only a fraction of this fraction will have bothered to rate it. In other words, the dataset is sparse, because only a small percentage of the possible entries are known (remember we talked about the graph of the web being sparse in chapter 5). Still, the total number of user-movie rating pairs entered in the database is several billion.

We can represent this input visually as a table, with users along the rows and movies along the columns, as in the left side of illustration 7.2. For each element in the table, the location indicates (i) which user and (ii) which movie, and the entry itself indicates (iii) the star rating. A question mark (?) is shown for the unknown ratings.

What are these inputs used for? They are used to adjust the different parameters of the prediction algorithm in the system. You can think of a parameter as a tunable knob or button, where changing the value of the parameter (i.e., turning the knob) will have some effect on what the system returns as its outputs. The input data is used during the training phase of the system, to set the parameters to values that are expected to yield good-quality outputs.

The Output

What does a recommendation system give us as output? The output is, first of all, a set of predictions for the ratings that a user would give to the movies she hasn't watched yet. This is shown on the right side of illustration 7.2, where the question marks from the left side have been replaced with some predicted value. You see these outputs have decimals (i.e., they are not just whole numbers between 1 and 5). How can we interpret these? Well, a prediction of, say, 4.2 is somewhere between 4 and 5, but closer to 4. It is 20% of the way toward 5, and 80% toward 4. So we can interpret 4.2 as

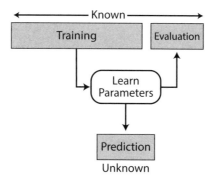

Illustration 7.3 For evaluation, the known data (i.e., the ratings that we have in the database) is split into two groups: one for training and one for evaluation.

saying there is a 20% chance that the user would give the movie 5 stars and an 80% chance she would give it 4 stars.

These predicted ratings are not what is actually shown to the viewers in the end, though. The final output of the system is a series of short, rank-ordered lists of movies recommended to individual users. How can we use the predictions to determine what the list will be for each user? We could apply several different criteria; for example, we could recommend five movies with highest predicted ratings that a user hasn't seen yet, or all the movies with a prediction of more than 4 stars.

Evaluating the Predictions

Now, how do we determine the quality, or performance, of such a system? The true test would be how many users actually like the movies that are recommended to them. But this information is hard to collect. Instead, we need some metric that can be used as a proxy.

What if in addition to making predictions on ratings that we don't know, we made some on ratings that we do know? Then we could see how well our predictions on these known ratings match the actual values. The closer they match, the higher quality we can expect in making predictions on the unknown ratings. This line of thinking is commonly used to evaluate these types of systems: (i) use some data to tune the parameters of the system, (ii) use the system to make predictions about target values (some known and some unknown), and (iii) compare the predicted values of the known targets to their actual values to get an expectation of quality.

To do this properly, we have to make sure that we separate the data that we use for training and the data we use for evaluation, as shown in illustration 7.3. Both of these sets are known, but they must be distinct, because we need to test the ability of the algorithm to predict new ratings it hasn't seen before. In other words, during the training phase, we need to withhold (or hold out, as it's known) the evaluation ratings from the data that we input to the system.

Once we have predicted and actual ratings to compare, how do we determine the quality? A standard way is the **root mean square error**, or RMSE for short. Its definition is a mouthful: you find the errors on all points in the dataset, square them, find the mean (i.e., average), and take the square root. We can also look at the MSE, which saves us the last (square root) step. Take the numbers 1, 2, and 3 as errors on three values, for example. Squaring them gives $1^2 = 1 \times 1 = 1$, $2^2 = 2 \times 2 = 4$, and $3^2 = 3 \times 3 = 9$. The average is

$$\frac{1 + 4 + 9}{3} = 4.67$$

which is the MSE. The RMSE then takes the square root: $\sqrt{4.67} = 2.16$. The lower the RMSE (or MSE) is, the higher we expect the quality of the system outputs to be.

Before diving into rating prediction algorithms in more detail, let's see how they played an important role in Netflix history.

The Netflix Prize Competition

Netflix's original prediction algorithm was called CineMatch. Recognizing how important it was to have the best rating prediction algorithm possible, Netflix launched a challenge in October 2006. Their open, online, international competition, called the **Netflix Prize**, awarded $1 million to the team that could develop an algorithm that would improve RMSE by 10% over CineMatch.

At the beginning of the competition, Netflix released a set of more than 100 million ratings to the public, as part of its records from 1999 to 2005. That amount of data could fit in the memory of standard desktops in 2006, making it easy for anyone in the world to participate in the competition. The rating data came from more than 480,000 users and 17,770 movies. On average, each movie was rated by more than 5,000 users, and each user had rated more than 200 movies.

This dataset is what the participants of the competition would have to train their algorithms with. Did it have ample information to make predictions for each user? At first glance, it would seem so. Digging further, it turned out that only a few users were responsible for pulling this average up to 200, by rating a large number of movies (one user had rated more than 17,000 movies!). For the majority of users, only a few ratings were available, posing an interesting challenge to aggregating users' individual preferences.

Netflix held back a few million ratings from the training data that only they had access to. The test set of 1.4 million ratings would serve as the final evaluation to determine the winner. CineMatch gave an RMSE of 0.9525 on the test set, which made the goal to drive the RMSE below $0.9 \times 0.9525 = 0.8573$ on this set. This may not seem too much different, but lowering the RMSE by even 0.01 can actually make a significant difference in the final recommendations.

Overall, the competition sparked some of the most intense activity that has been seen in research on recommender systems in recent years. More than 5,000 teams worldwide entered more than 44,000 submissions. Cine-Match was beaten within a week of the start of the competition in October 2006, but it would be almost 3 years until the first team achieved more than a 10% improvement in June 2009. In the end, the top two teams—The Ensemble and BellKor's Pragmatic Chaos—achieved the same improvement in RMSE of 10.06% on the test set. Since the latter team had submitted their algorithm 20 minutes earlier, they were declared the winner.

For more details on the timeline of this competition and on how the datasets were partitioned for training and evaluation, check out Q7.1 and Q7.2 on the book's website.

BUILDING THE BASELINE PREDICTOR

Obtaining the last few percent improvement in RMSE on Netflix's dataset requires many algorithms blended together and thousands of model parameters tuned just right. It is not our intention to explain the details involved in such procedures. Instead, we focus on two of the first steps: the baseline predictor and the neighborhood model, simplifying the math involved along the way as much as possible. As you will see, at the inner workings of these schemes is the notion that we can leverage information

	I	II	III	IV	V
A	5	?	4	?	4
B	4	3	5	**3**	4
C	**4**	2	?	?	3
D	2	**2**	3	1	2
E	4	?	**5**	4	5
F	4	2	5	4	**4**

Illustration 7.4 Example of a dataset with six users (A–F) and five movies (I–V).

from within the "crowd" of data we have already to infer user preferences and movie characteristics for making predictions.

Our Example Dataset

Take a look at the example dataset in illustration 7.4. It consists of six users (A–F), one per row, and five movies (I–V), one per column. The entries that are neither bold nor question marks will be used as the training data for us to base our predictor on. The five bold entries will serve as the test data, so that we can evaluate our algorithm's performance on this set. The five question marks are the final outputs for the system to predict in the end.

Clearly, this example is nowhere near the scale of Netflix's dataset with tens of millions of users and tens of thousands of movies. Also, Netflix's data is much more sparse: in our example, we have 83% of the entries in the table filled, whereas Netflix has only fractions of 1% of all possible ratings. Still, this small dataset will suffice to get the main ideas across.

Naive Predictor

How to develop a predictor? For starters, we could try to take the average of all the entries in the training set and attribute this to all the unknown entries. For the 20 numbers in the training data, this is

$$\frac{5 + 4 + 2 + 4 + 4 + 3 + 2 + 2 + \cdots}{20} = 3.5$$

which you can see on the right side of illustration 7.5.

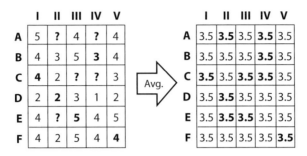

Illustration 7.5 Original ratings (left) and naive predictor (right).

Let's evaluate the quality of this predictor by calculating the MSE, which we can just call the "error" for short. As mentioned earlier, we must take the average of the "squared differences." We will focus here on the test set, but we can apply the same steps on the training set.

First, for each element in the test set, we find the difference between the raw rating and the predicted rating, and square it. Take user B and movie IV, for example: the prediction is 3.5, and the raw rating is 3, so the squared difference is $(3 - 3.5)^2 = 0.25$. How about for user D, movie II? It's $(2 - 3.5)^2 = 2.25$.

Once we have obtained the squared differences for each point in the test set from illustration 7.5, we find the average of all these values:

$$\frac{(4 - 3.5)^2 + (2 - 3.5)^2 + (5 - 3.5)^2 + (3 - 3.5)^2 + (4 - 3.5)^2}{5}$$

$$= \frac{0.25 + 2.25 + 2.25 + 0.25 + 0.25}{5}$$

$$= 1.050$$

On the training data, you can find the error to be 1.350.

Baseline Predictor

Relying on the average overall rating in Netflix's dataset would be rather naive. It is like using the same average customer review for every product on Amazon in chapter 6. Maybe we can incorporate two things that we learned back in that chapter: some reviewers tend to be easier/harsher critics, and some items (in this case, movies) are naturally better/worse than others.

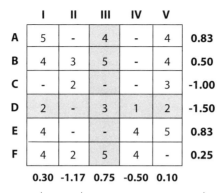

	I	II	III	IV	V	
A	5	-	4	-	4	**0.83**
B	4	3	5	-	4	**0.50**
C	-	2	-	-	3	**-1.00**
D	2	-	3	1	2	**-1.50**
E	4	-	-	4	5	**0.83**
F	4	2	5	4	-	**0.25**
	0.30	**-1.17**	**0.75**	**-0.50**	**0.10**	

Illustration 7.6 The user and movie bias terms are given at the end of their respective rows/columns.

For instance, take a look at user D. The highest rating she gave was a 3, and this was to a movie (III) that received all 4s and 5s from other users. She also gave two 2s, and was actually the only person to give a 1. User D therefore seems like a harsh critic, and the value we predict for her rating of movie II should reflect this. Also, take a look at movie III: with the exception of the one 3 it received (which was by the harsh critic D), it received all 4s and 5s. We expect this movie is more likely to be well received by those who haven't rated it yet.

In other words, each user and each movie has its own rating bias. The **baseline predictor** posits that the rating for a specific user-movie pair will be offset from the overall average by the corresponding biases, that is,

$$\text{Rating} = \text{Average} + \text{User's bias} + \text{Movie's bias}$$

We already know how to find the (overall) average. What about the bias terms? We can figure them out by considering the user-movie interactions (i.e., how a given user rated movies, and how a given movie was rated by users). In general, to find their best possible values, we would need to solve an optimization problem. Rather than doing that, let's take an intuitive approach: for a user, find his/her average rating across all movies he critiqued (in the training data) and compare this to the overall average. If it's higher, this is an indication of how lenient he/she is relative to the whole dataset, and if it's lower, it gives us an idea of how much more critical he/she is. Similarly, for each movie, we find its average over all the users that critiqued it and compare that to the overall average.

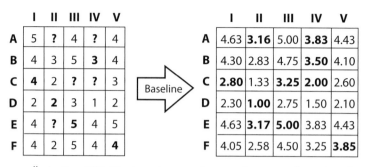

Illustration 7.7 The baseline predictor is shown on the right side.

Refer back to illustration 7.4. For the harsh user D, there are four ratings in the training data: 2, 3, 1, and 2. Therefore,

$$\text{Bias D} = \frac{2+3+1+2}{4} - 3.5 = -1.5$$

This is significantly less than zero, as we would expect: her average rating is much less than the overall average. What about for the good movie III? There are also four ratings: 4, 5, 3, and 5. Therefore,

$$\text{Bias III} = \frac{4+5+3+5}{4} - 3.5 = 0.75$$

This is greater than zero, as we would expect. You can find the remaining bias terms in the same way. The values are in illustration 7.6, at the ends of the rows (for the users) and columns (for the movies).

With these terms in hand, we can make the baseline predictions. What would we get for user D rating movie III?

$$\text{Average} + \text{Bias D} + \text{Bias III} = 3.5 - 1.5 + 0.75 = 2.75$$

This is only 0.25 away from user D's actual rating. Taking one of the ratings that we don't have, what would we get for user A, movie II?

$$\text{Average} + \text{Bias A} + \text{Bias II} = 3.5 + 0.83 - 1.17 = 3.16$$

You can repeat this for each of the 30 user-movie pairs. The full baseline predictor is shown on the right side of illustration 7.7. One thing you'll notice is that none of the predictions goes below 1, and none of them goes above 5. How come? When predictions go outside this range (e.g., for user E, movie III, $3.5 + 0.83 + 0.75 = 5.08$), we can only make the error worse by keeping them there, because it is not possible for the actual ratings to go above 5 or below 1. We should always restrict the predicted ratings to be in this range.

So what is the error when using the baseline predictor? Comparing the predicted to actual ratings in illustration 7.7 for the test set, we have

$$\frac{(4 - 2.80)^2 + (2 - 1.00)^2 + (5 - 5.00)^2 + (3 - 3.50)^2 + (4 - 3.85)^2}{5}$$
$$= \frac{1.44 + 1.00 + 0.00 + 0.25 + 0.023}{5}$$
$$= 0.543$$

Compared to the naive predictor's test set error of 1.050, we have made an improvement of $(1 - 0.543 / 1.050) \times 100\% = 48\%$. You can find the baseline predictor's error on the training data to be 0.223, which is an improvement of 83% from 1.350. Not too shabby!

HELP FROM "NEIGHBORS"

So far, we have been taking averages along single rows or columns of the rating data, to find user-movie interactions. What about leveraging similarities among different movies and between different users? This is the essence of the **neighborhood model**, where we call two users "neighbors" if they share particularly similar (or dissimilar) opinions about movies, and two movies "neighbors" if they were rated particularly similarly (or dissimilarly) by users. Neighborhood modeling is one of the more intuitive methods in **collaborative filtering**, where we filter a dataset for patterns by looking at how entities "collaborate" (in this case, how they rate or are rated). We turn to this method next.

Similarity and Dissimilarity

Let's say Anna and Ben both like the movies "Good Will Hunting" and "A Beautiful Mind," whereas they both do not like "Lion King" and "Aladdin," as in illustration 7.8. In this case, Anna and Ben seem to be **positively correlated** users (i.e., they have similar preferences). So if we know Anna likes

Similar Users		Good Will Hunting	A Beautiful Mind	Lion King	Aladdin	Jurassic Park
	Anna	👍	👍	👎	👎	👍
	Ben	👍	👍	👎	👎	?

Illustration 7.8 Two users are similar when they tend to have the same opinions on movies.

"Jurassic Park," we would expect Ben to like it, too, and if we know Anna does *not* like it, we would expect Ben not to like it either. Correlation also works in reverse: if Ben did *not* like the first two movies, and he liked the second two, then Anna and Ben would seem to be highly **negatively correlated**. In this case, if Anna likes "Jurassic Park," we would expect Ben not to like it, and vice versa.

Take another example. Suppose "Good Will Hunting" and "A Beautiful Mind" are both rated high by Anna and Ben, and they are both rated low by Charlie. Then, if Dana rated "Good Will Hunting" high, we would expect her also to rate "A Beautiful Mind" high, since the others' opinions have indicated that the movies are positively correlated (i.e., similar in type, quality, etc.). Similarly, if Dana rated one of the movies low, we would expect her to rate the other one low as well. This also works in reverse, as you can see in illustration 7.9: if Anna, Ben, and Charlie had rated "A Beautiful Mind" opposite to what they rated "Good Will Hunting," then the movies would be negatively correlated, and we would expect Dana to rate the movies opposite as well.

How do we quantify this notion of similarity? The standard metric is called the **cosine similarity**. Calculating this requires a bit of geometry, so we won't go into the exact math here; if you're interested in the details, check out Q7.3 on the book's website. The computed movie-to-movie similarities are given in illustration 7.10. From this table, we can say, for example, that the similarity between movies III and IV is 0.50.

How do we interpret these values? Here are a few important properties of cosine similarity that will help us:

- It will always lie between −1 and +1.
- A perfect positive correlation has the value +1. A value close to +1 is a strongly positive correlation (i.e., there is a high degree of similarity).

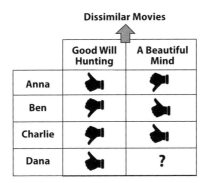

Illustration 7.9 Two movies are dissimilar when they tend to receive opposite feedback from users.

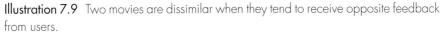

	I	II	III	IV	V
I	---	-0.11	-0.82	0.01	-0.74
II	-0.11	---	-0.74	-1.00	0.88
III	-0.82	-0.74	---	0.50	0.79
IV	0.01	-1.00	0.50	---	0.48
V	-0.74	0.88	0.79	0.48	---

Illustration 7.10 Table of movie-movie similarities. In each movie's column, its closest neighbor is highlighted.

- A perfect negative correlation has the value −1. A value close to −1 is a strongly negative correlation (i.e., there's a high degree of dissimilarity).
- A complete lack of correlation has the value 0. A value close to 0 means there is only a weak correlation (i.e., neither similar nor dissimilar).

What can we surmise about, say, movies III and V then? Because 0.79 is close to +1, these two movies are positively correlated. This makes sense from the ratings in illustration 7.4: users A and B each rated both of these movies relatively high, and user D rated both of these movies relatively low (we can't consider the other users, because they have not rated both movies

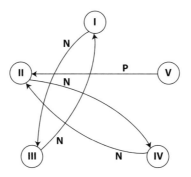

Illustration 7.11 Graph of nearest-neighbor movie relations. The link from a node indicates its closest neighbor. ``N'' means the correlation is negative, and ``P'' means it's positive.

in the training set). What about movies I and II? Because −0.11 is close to neither +1 nor −1, this is saying that they are not correlated one way or another.

Choose Your Neighbors Wisely

These movie-to-movie similarity values—or, alternatively, the user-to-user similarities—are what we use in the development of the neighborhood model. So how do we go from "similarity" to what constitutes a "neighbor?" We could use various rules to determine the neighbors of a particular movie. For instance, we could go down each column and select the three movies with the highest similarity values. Or we could say that any movie is a neighbor if it has a similarity value higher than some threshold. When we say "higher" here, we mean higher in absolute terms: it could be either a strong positive or a strong negative correlation. Both types can be useful.

To make the subsequent math simpler, let us choose the best single neighbor for each movie. You can see the result in illustration 7.10: a highlighted box indicates that the movie in the column will select the movie in the row as its neighbor. For example, V chooses II as its neighbor. Does that mean that II will choose V? No: even though the similarity values are **symmetric** (i.e., the similarity of II to V is the same as that of V to II), neighbor selection doesn't have to be. II's correlation with IV is larger than with V, so II chooses IV as its neighbor instead.

We can represent the neighbor selections as a graph, too. You can see that in illustration 7.11: each movie is a node, pointing to its chosen

	I	II	III	IV	V
A	0.37	**?**	-1.00	**?**	-0.43
B	-0.30	0.17	0.25	---	-0.10
C	---	0.67	**?**	**?**	0.40
D	-0.30	---	0.25	-0.50	-0.10
E	-0.63	**?**	---	0.17	0.57
F	-0.05	-0.58	0.50	0.75	---

Illustration 7.12 Table of errors for the baseline predictor.

neighbor. If we allowed movies to choose more neighbors, each node in this graph would have more outgoing links.

Now, what to do with these neighbors? We want to use them in such a way that we expect will improve the quality of our predictor. To do this, we work with the errors that were obtained by the baseline predictor in illustration 7.7, hoping to have this method calibrate for the errors that may have carried over from the training to the test set. We get the error for each individual user-movie pair by subtracting the baseline prediction from the raw rating, as in illustration 7.12. These are pairwise errors (i.e., one for each user-movie pair) rather than a single summary error like MSE which measures overall quality.

Now we have all we need to apply the neighborhood method. Let's take movie V, user C as an example first. Movie V's closest neighbor is II, which has a high positive correlation (+0.88). User C rated movie II, and the predictor made an error of +0.67 in that rating. What does this error tell us? Our prediction was too low by 0.67, and because V is similar to II, our prediction of user C's rating of movie V may have been too low as well. So we add 0.67 to the baseline prediction of 2.60:

$$2.60 + 0.67 = 3.27$$

Compared to the actual rating of 3, the error here is smaller: we are too high by 0.27 rather than too low by 0.40.

Let's take another example: user B on movie IV. Movie IV's closest neighbor is II, which has a perfectly negative correlation (-1.00). The error in predicting user B's rating of movie II was $+0.17$, telling us we were too low by 0.17. Because IV is dissimilar to II, our prediction of user B's rating of movie IV may have been too high. So what to do in this case? We subtract 0.17 from the baseline prediction of 3.50:

$$3.50 - 0.17 = 3.33$$

Again, the error here is smaller (0.33 rather than 0.5).

So, if the correlation is positive, then we add the neighbor's baseline error, and if it's negative, then we subtract the error:

$$\text{Rating} = \text{Baseline} \pm \text{Neighbor's error}$$

This is a simple form of a **neighborhood predictor**, where we only use the closest neighbor. As mentioned, this method can be extended to using more neighbors, but we won't get into that here.

We can view what the neighborhood predictor is doing as a type of negative feedback that we talked about in part I. It is using the errors made by the baseline as a type of "feedback signal," correcting for potential "mistakes" in the output, like how power control adjusts for differences in the target signal quality.

The full set of predictions is shown on the right side of illustration 7.13. Some of the values are the same as in the baseline predictor shown in illustration 7.7; when would this happen? When either the user did not rate the neighboring movie, or the neighboring rating is not part of the training data. For instance, take user C's rating of movie I. Movie I's neighbor is III, and C did not rate that one. So we don't change the baseline prediction.

We can't expect the error to get lower for each individual rating. As an example, take user F's rating of movie V: the neighborhood method predicts that to be 3.27, which is further away from the true rating of 4 than the baseline predictor's 3.85 was! But even though the neighborhood predictor may be off in some individual cases, we expect that it will help more than it will hurt overall. To check that, let's find the overall error. For the

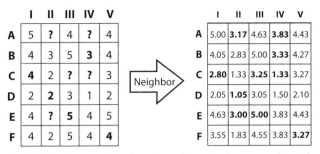

Illustration 7.13 The neighborhood predictor is shown on the right side.

test set, we get

$$\frac{(4 - 2.80)^2 + (2 - 1.50)^2 + (5 - 5.00)^2 + (3 - 3.33)^2 + (4 - 3.27)^2}{5}$$
$$= \frac{1.44 + 0.25 + 0.00 + 0.11 + 0.53}{5}$$
$$= 0.466$$

Similarly, we can evaluate the error on the training data to be 0.134. These are improvements of 56% over the original, naive average predictor on the test set and of 90% on the training set. Compared with the baseline predictor, the improvements are 14% and 39%, respectively.

To wrap things up, you can see a comparison between the errors of various predictors obtained on the training and test data for our example in illustration 7.14. The first three sets of bars are the results for those methods we walked through in this chapter. The fourth shows what error we would have obtained if we had used the two closest neighbors for each movie, rather than just the single closest. As it turns out, the error on the training set stays roughly the same, while on the test data it rises significantly, so one neighbor was actually a better choice in this case.

We said earlier that computing the bias values in the baseline predictor based on the column/row averages was not the optimal approach. The fifth set of bars shows the error that we would have obtained if we had solved an optimization problem for the bias terms: it lowers the error from what we had in the second column. Finally, the errors that would have been obtained if we had used the neighborhood methods with the optimized baseline are given. As it turns out, in hindsight, using two neighbors would have been the better choice: that is what leads to the biggest improvement in this example, out of the methods in illustration 7.14.

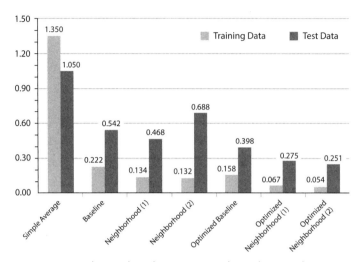

Illustration 7.14 Errors obtained on the training and test data used in our example for different predictors.

It wasn't our objective in this chapter to explain all the details of the winning algorithm in the Netflix Prize competition. That requires a lot of advanced mathematical techniques. But you now know the basic ideas behind how Netflix makes movie recommendations: by finding patterns among users and among movies and using these patterns to make predictions of unknown opinions.

8

Learning Socially

We've just seen how decisions can be made based on information gathered from "crowds." In this chapter, we turn to situations where people learn from one another, each bringing their own wisdom to the crowd.

TECHNOLOGY AND PEDAGOGY: SCALING UP LEARNING

One motivation behind the creation of the Internet was to enable easier access to content. Today, in addition to surfing the web for specific chunks of information one at a time, we can enroll in entire courses filled with content organized around a particular subject. Rather than going to a physical classroom to participate in a lecture, the Internet has made it possible to watch videos of lectures from our computers.

Online learning, as it's called, has exploded in popularity. Many higher education institutions now offer full degree programs online, and most colleges and universities offer at least some parts of some courses online for their students. In fact, more than 25% of college students were taking at least one course online in the fall of 2012.

In the past decade, the Internet has scaled up learning in two senses. The total number of people taking online courses has become much larger, as has the number of people who may be taking the same class at the same time. In some cases, the latter has pushed into the hundreds of thousands! When the number of students in a class grows very large, it becomes challenging for the instructor(s) to answer all the questions and address all the needs that students will have. The students need to be able to learn socially, that is, to learn from and to teach one another.

Learning from a Distance

Saying that **distance learning** started with the Internet would be like saying that transportation began with the automobile. Most communication

	Auditory	Visual	Textual	Social	Synchronous
In Class	✔	✔	✔	✔	✔
Mail		✔	✔		
Radio	✔				
Television	✔	✔			
Internet	✔	✔	✔	✔	

Illustration 8.1 Comparison between technologies that have been used for education, in terms of learning modes they can readily support.

systems have been used to facilitate education over the years in one way or another. A few degree programs operating by postal mail—in which materials and assignments would be mailed to peoples' homes—had emerged as early as the mid-1800s. When radio and television became popular in the early and mid-1900s, some universities began broadcasting course lectures over these types of networks too.

Online programs, online degrees, and even online universities began to spring up in the 1990s with the rise of the web. By 2003, 80% of colleges had at least one class that used online technology in some way. Fast-forward to 2014, and less than 5% of all public colleges and universities did not offer some form of online program.

What does online learning offer that prior distance learning did not? It's helpful to compare by which modes of instruction each technology supports. A basic set is shown in illustration 8.1: hearing the instructor (auditory), seeing the information as it is being written (visual), reading the material (textual), discussing with peers (social), and asking questions in a real-time (synchronous) fashion. As we've experienced in school, a face-to-face classroom supports all these modes. The Internet allows for all but synchronized learning, with prerecorded lecture videos (auditory and visual), forums for student discussions (social), and supplemental material (usually textual). Each of the other technologies can only support one or two modes.

Different types of online learning have emerged, too. Some courses are open for anyone to sign up, whereas others are to be taken as part of degree programs. Some have massive enrollment numbers, while others are on the order of traditional classroom sizes.

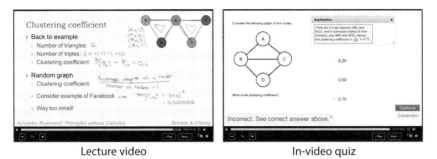

| Lecture video | In-video quiz |

Illustration 8.2 Example of a lecture video (left) and an embedded in-video quiz question (right), taken from an online course based on this book (you'll see this material when we get to chapter 14).

The most recent type of online learning traces back to 2002, when MIT created an online resource of free course materials from the university that anyone could access. Many were puzzled by this bold and apparently unnecessary move. Why would MIT decide to put course material online for free? What would MIT get in return for disseminating knowledge so widely? It turns out that this was an inspiring pathfinder that became a major trend to transform classroom-sized online courses into **massive open online courses**, or MOOCs. Many top-tier universities in the United States joined the movement in the decade and a half following MIT's announcement, partnering with MOOC sites like Coursera, edX, Udacity, Udemy, and others to deliver their courses. Actually, the predecessor course of this book was the first MOOC on networking back in 2012.

The "MOO" in MOOC

There are now more than a dozen MOOC providers. Each possesses certain operational differences, but most of them have a few properties in common. By virtue of the name "MOOC providers," their courses are delivered online and keep enrollment open to anyone, either for free or at a cheap price. The most popular way for instructors to teach a MOOC is through YouTube-style videos of lectures with quiz questions embedded in them, as you can see in illustration 8.2. These platforms also incorporate social networking through discussion forums, where students can ask and answer one another's questions.

As a by-product of open, online delivery, MOOCs have attracted massive enrollments. How "massive" exactly? A standard MOOC offering will have tens, and up to hundreds of thousands of students, from all around the globe. Imagine being in a face-to-face classroom with that many people! While it hardly seems possible in person, advances in web technology have made it feasible to bring this many people together in virtual classes online.

Even in traditional classrooms with only a few dozen students, any teacher will have her hands full trying to address all the questions, concerns, and misunderstandings of each person. As the instructor, she has the important job of **individualizing learning** (i.e., making some adjustments to how she teaches the material for each student). Achieving this in a MOOC is complicated. The massive scale makes each teacher responsible for many more students. Open enrollment will tend to attract a set of students that is more heterogeneous with respect to their background knowledge and expectations for what they will get out of the course, calling for a broader range of differentiation in the first place. In addition, it is harder for an instructor to figure out each student's learning needs when they are only interacting online.

To make a long story short (but feel free to check out Q8.1 to Q8.6 on the book's site for more information), scaling traditional teaching methods to the size of MOOCs is challenging. A **one-size-fits-all** instructional style can be applied to everyone, but even if strategically designed based on the "average" learning need in the course, it is not likely to work well for many students (illustration 8.3). There is one component of online education that is scalable, though: the learning that occurs in social networks, as we'll see next.

LEARNING SOCIALLY

What can Bob do when he is confused by part of the material in the course he is taking? If he is in the presence of a large body of peers, it is unlikely that the instructor can address all of Bob's specific learning needs along with everyone else's. On the upside, it is also more likely that at least one of his peers will be able to help him. There are probably many others who would benefit from seeing his question answered, too.

What we are describing here is the process of **social learning**, where students learn through interaction and collaboration with one another.

Illustration 8.3 If an instructor had to choose one way to teach to everyone, her best bet would be to base the teaching on the ``midpoint'' of the learning needs in each case. In a classroom, this is still close to the majority of students, but in a massive open online course (MOOC), many students are far from it.

They can work together, discuss what they have learned, and answer one another's questions. Social learning is a critical component of MOOC, and of education in general. The crowd gains its wisdom not from the knowledge of one, but from the collective knowledge of many.

Discussing on the Forums

What is the vehicle for social learning in online courses? The primary means for student-to-student (and teacher-to-student) interaction is the **discussion forum**. You can see a snapshot of a discussion forum in illustration 8.4: people communicate through sequences of messages, where each "message" is either a post or a comment on a post. Messages are made up of some text written by a single student. A series of posts (and their comments) will be contained in a larger **thread**, the entire forum being a collection of such threads.

How do students use the forums? Let's return to Bob's question. He can start by checking to see whether someone else has already posted the same question on the forums, or one similar enough that its answer would suffice. He could look for this by entering phrases similar to his question into the search bar, or by browsing through the forum manually.

What if Bob does find a post that is already asking his question? Then he could see if an answer is in one of the comments or other posts there. If it is, then he could up-vote the answer if it's good, or down-vote it if it's bad, to give feedback. If it hasn't been answered, then he could up-vote the question, so that someone else may notice it and answer it quicker.

What if he finds no such post? Then he could look to see whether there is a similar thread of discussion and write a new post in it. If he doesn't find any thread that is close to what he wants, he could create a completely new

Illustration 8.4 A massive open online course discussion forum is typically broken down into a series of threads (left). In each thread is a series of posts, and each post can have comments written in response to them (right).

thread with an appropriate title. You can find a flowchart of this process in Q8.7 on the book's website.

Social Learning Networks (SLNs)

Peer-based learning creates a **Social Learning Network**, or **SLN** for short. The three main characteristics of an SLN are contained in its name:

- *Learning* is the process taking place. It represents the acquisition of knowledge about the subject matter of interest. Typically, this can be broken down into a set of topics that make up the material.
- *Social* learning hinges on interaction among peers. There may be designated teachers/instructors, but without collaboration between the individual learners, the network would not be effective.
- The social *network* among peers depends on, and in turn influences, the learning process.

SLN gives an interesting spin on how individualization can be achieved. Rather than relying on instructors to respond to individual needs that arise, we can search for solutions inside the SLN. The learning process becomes scalable, because more students should lead both to more questions and to more potential answerers of the questions, preserving the balance between the two. You can see this idea in illustration 8.5.

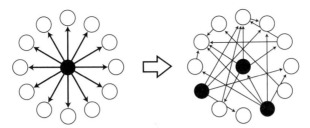

Illustration 8.5 When the instructor (center node) is required to address all of the needs in a course (left), the learning process is not scalable. The premise behind a social learning network is that learners themselves can also act as teachers (right).

Similar to how Netflix captures the movie ratings of its subscribers and Amazon collects the product reviews of its shoppers, MOOC platforms can record data about students as they proceed through courses. This includes both how they interact with one another (e.g., through the posts, comments, and votes made in a discussion forum) and how they digest the course content (e.g., through the sequence of clicks they make while watching a lecture video). The question then becomes how to analyze and leverage this data.

Where Are SLNs?

We have discussed SLNs in the context of online education and MOOCs so far. Where else would we find these networks? SLNs exist in any scenario by which learning occurs socially. Take the **flipped classroom**, where watching the lectures that are usually taught in class becomes part of the homework, and class time is instead used for discussion and interaction. When education technology is deployed in conjunction with these courses, data about SLNs can be collected.

Applications exist outside education, too. Rather than students in classrooms, consider employees of corporations. When employees are first hired, they typically have to go through an onboarding process in which they learn additional material and skills needed to perform their jobs. They also frequently take corporate training courses throughout their careers. To facilitate social learning in these classes and beyond, corporations typically have **enterprise social network** platforms like Jive and Yammer.

Finally, consider **Question and Answer (Q&A) sites** like Quora, Yahoo! Answers, and Stack Overflow. Functionally similar to the discussion forums

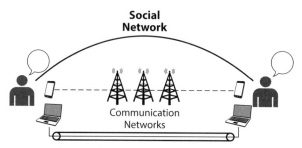

Illustration 8.6 For social networks, we are interested in the interactions among people, whereas in studying technological networks, we are focused on the communications and channels among people's devices.

in online courses, Q&A sites create SLNs among question askers and question answerers. They also have incentive structures in place to encourage constructive participation, with users gaining points when they receive upvotes on questions or answers.

In the rest of this chapter, we outline a few of the important areas of investigation regarding these networks. As one of the newest topics at the time of this writing, you will see many interesting not-yet-answered questions raised that are motivating continued investigations. Just as important, you will see examples of constructing networks that visualize relationships.

VISUALIZING AN SLN EFFECTIVELY

Think of a series of discussions in a forum, as in illustration 8.4. There's a lot of information we can capture about an SLN here. Who posts in a thread? Who responds? What are the contents of the posts and the responses? And so on. How can we visualize the network from this information? We can't put all the data in a single picture, because it would be too cluttered and convoluted to draw meaningful insights from. We have to think carefully about what information is important for us to depict.

How do we represent networks? With graphs, like the webgraphs from chapter 5. In the graph of an SLN, what should we have as the nodes? What should the links represent? Should they be directed (from one node to another) or undirected? Should they be weighted (with some number that gives a magnitude to each link) or unweighted?

Perhaps the place to start is to first ask a more fundamental question. How is an SLN different from the other types of networks we have looked at so far (e.g., a network of webpages as in chapter 5 or a network of devices as in chapters 1 and 2)? For one, an SLN is a type of social network among people, as opposed to a communication network among Internet-connected devices, as shown in illustration 8.6. The medium of communication between people in an SLN may still be the Internet, but in this case we are less concerned with how the information is physically transferred between peers than what information is being shared and which social connections are created as a result. We will look at other types of social networks in chapters 10 and 14.

So the graph of discussions should depict the process of information sharing. Still, there are many variations for what we could choose as the nodes and links, which depend on our ultimate goal in the visualization.

Graphs of Students

The most obvious choice for nodes is to have them represent different students. Then what would a link between two students—say, Alice and Bob—indicate? It should be representative of some relationship between them, with respect to how they've shared information. Four possibilities are:

(a) Whether Alice and Bob have participated in a discussion together.
(b) How many times Alice and Bob have participated in a discussion together.
(c) Whether Alice has responded to a post made by Bob, and vice versa.
(d) How many times Alice has responded to a post made by Bob, and vice versa.

The differences between these statements may appear subtle, but they each invoke different types of links. Are the links directed or undirected, and weighted or unweighted, in each case? You can see these different types in illustration 8.7.

In (a), we have an undirected, unweighted graph: if we find a thread that Alice and Bob both posted in, then they are connected. In (b), we add magnitude to (a) using a **co-participation** count, which makes an undirected, weighted graph: for example, if we find Alice and Bob both posted in three threads, then the weight on the link between them is 3.

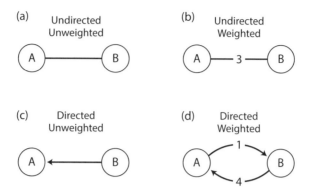

Illustration 8.7 Four different permutations on link properties that can be used to visualize a social learning network as a graph.

(c) is different from (a) because we are talking about responses. If Bob has answered a question written by Alice, that doesn't mean Alice has answered a question written by Bob. We need arrows to make this distinction, and so we have a directed, unweighted graph. In (d), we go a step further from (c) and count the number of responses, which gives us a directed, weighted graph: if Bob has responded to Alice four times, and Alice has responded to Bob once, then Bob's link to Alice has a weight of 4, and Alice's link to Bob has weight 1.

Let's walk through an example of visualizing the (small) discussion forum shown in illustration 8.8 in terms of these types of graphs. There are four students—Alice, Bob, Charlie, and Dana—each of whom have participated in the forum a different amount. In each of the four threads, a varying number of posts has been made—three in thread I, two in thread II, and so forth. Each post has a person asking a question (the top name listed) and a person responding to it (the bottom name). (In reality, there could be many responses given to one question, and a post may not even have a question to begin with.)

For each of the graph types, what will we have as the nodes? There will be four, one per student. Now, how about the links? Let's walk through each case separately, for Bob and Charlie:

(a) *Undirected, unweighted*: Have Bob and Charlie participated in a thread together? Yes, they both contribute to threads I and IV. So we draw a link between them in this graph.

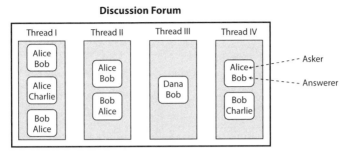

Illustration 8.8 Small discussion forum with four threads, four students, and eight posts.

(b) *Undirected, weighted*: How many threads have Bob and Charlie participated in together? Two. So their link weight is 2.

(c) *Directed, unweighted*: Has Bob responded to a post made by Charlie? No. Has Charlie responded to a post created by Bob? Yes, in thread IV. So Charlie links to Bob, but Bob does not link to Charlie.

(d) *Directed, weighted*: How many times did Bob respond to Charlie? Zero. How many times did Charlie respond to Bob? One. So the link from Charlie to Bob has weight 1.

After doing this for the rest of the student pairs (which we encourage you to do!), we get what is shown in illustration 8.9. In the directed cases (c and d), Alice and Bob are the only pair pointing to each other; Alice responded to Bob two times (once each in threads I and II) and Bob responded to Alice three times (once each in threads I, II, and IV).

Graphs of Students and Threads, Threads and Threads, …

Using students as nodes is helpful in conveying the "who" of the network (i.e., which students are interacting). But it does not explain the "what" (i.e., the topics being discussed in the conversations). This information can be important to depict as well, because it can give an indication as to, for example, what topics tend to be discussed the most or which tend to confuse people.

How can we discover what the key topics are? Raw discussion text tends to be too long and uninformative in its unprocessed form. What we can do is call on appropriate methods from **natural language processing** to extract the topics underlying the text. An example of what topic extraction could

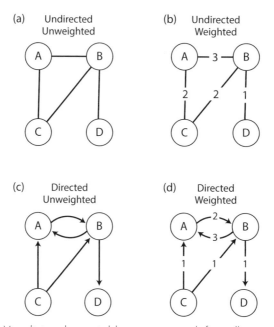

Illustration 8.9 Visualizing the social learning network from illustration 8.8 with four graph types.

output is shown in illustration 8.10: we get a set of topics, each comprised of certain keywords that tend to occur together. Each post can then be associated with one (or multiple) topics that are contained in its text.

How can we infer whether each student is tending to ask or answer questions on specific topics? This may seem a trivial undertaking: can't we just check whether the student's post on a topic contains a question mark in it or not? Unfortunately, it's not that easy, because there are many counterexamples, like asking a question without using a question mark (e.g., "please explain to me how this works"), providing an answer in the form of a question (e.g., "you would think they are different, wouldn't you?"), or just not punctuating properly.

Topic extraction and question detection are both interesting areas of research in **information retrieval**. Rest assured that high-quality procedures for these functions do exist today. We are going to suppose moving forward that (i) each thread is distinct enough to represent a separate topic, and (ii) we have accurately determined which posts/comments are questions and which are answers.

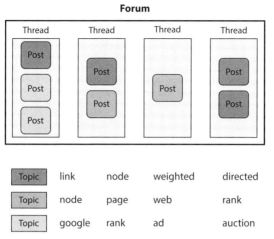

Illustration 8.10   Example of what could be obtained when topics are extracted from a discussion forum. Topics are made up of words that are found to occur together, and each post is associated with its most prominent topic.

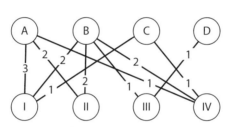

Illustration 8.11   Representation of the discussion forum from illustration 8.8 as a bipartite graph connecting students and topics.

How can we visualize the relationship between the students and the topics? We can augment our visuals with a second type of node, one per topic, rather than just depicting students in the graph. Going back to our example in illustration 8.8, we will end up with something like illustration 8.11. Here, the student nodes are on the top, the topic (thread) nodes are on the bottom, and each link indicates the number of times a student participated in a topic.

Illustration 8.11 is a **bipartite graph**: it has two separate sets of nodes (here, students and topics), with each link connecting a node in one of the sets (i.e., a student) to a node in the other set (i.e., a topic). This particular graph is weighted, but if we are only interested in whether a student

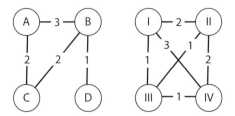

Illustration 8.12   Student-to-student (left) and thread-to-thread (right) co-participation graphs obtained from the bipartite graph in illustration 8.11.

discussed a topic or not, we could drop the magnitudes from the links to get an unweighted bipartite graph.

We have a lot of useful information in this illustration. We can see which topics each student discussed, which students discussed each topic, and the frequencies on top of that. One thing it hides is the nature of each student's posts: are they questions or answers? For example, we can see that Bob has contributed to each topic, in a total of seven posts, but we would have to refer back to illustration 8.8 to see that the split between asking and answering was roughly half: three in the former, four in the latter. Loss of detail is the price we pay in our search for concise visual representations of networks.

On the other hand, depending on what insights we are looking for, illustration 8.11 might contain too much information detracting from key messages. What if we want to see how "similar" two students tend to be in the topics they discuss, or how "similar" two topics tend to be in the students that have contributed to them? One way to figure these things out is through co-participation counts in the bipartite graph. For students, we consider each pair, counting the number of threads that both have posted in. Take Bob and Dana, for example: thread III is the only place they both posted together, so their co-participation is 1. For threads, we also consider each pair, this time counting the number of students that posted in both. Take threads II and IV: Alice and Bob each posted in both, so the co-participation is 2.

Repeating this for each of the other pairs, we get the graphs in illustration 8.12. Here, it is much easier to see, for example, that threads I and IV have three of the same students posting. But we can no longer see which three students were posting in both I and IV; for that, we'd have to refer back to the bipartite graph to see they were Alice, Bob, and Charlie. The

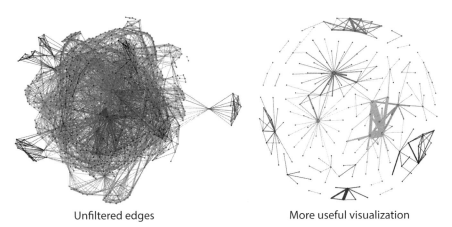

Unfiltered edges   More useful visualization

Illustration 8.13 Representing the social learning network of students in one of our massive open online courses.

student-to-student network here is the same as in the unweighted, directed graph from illustration 8.9, because we are using co-participations in each case.

Dividing a bipartite graph into two separate graphs, one for each node set, is known as **network projection**. Projection types are distinguished by the rules that they use to come up with weights between the nodes in the resultant graphs. The one taken here is known as a simple weighting, where each link in the bipartite graph is equally weighted.

PUTTING SLN RESEARCH TO THE TEST

So now we know a few types of graphs we can use to visualize different aspects of an SLN. But the size of the network in the example we used is orders of magnitude smaller than what we would find in a real MOOC discussion forum. How would these graphs hold up in practice?

Real-World SLNs

We'll take a quick look at the graphs from the discussions in one of our own MOOCs. In illustration 8.13, we have used the undirected, unweighted, student-to-student approach for visualizing the resulting SLN.

Take a look first at the left graph. A link between two students indicates they participated together in at least one thread. Although the graph may be aesthetically pleasing, it is really rather difficult to draw any meaningful conclusions from it, because of the sheer number of nodes and links. Plus, we are only showing the 713 students here who posted in the forum more than once; otherwise, the graph would be larger yet.

To uncover useful information about the SLN, we need to "clean up" this visual. What are the possible ways to do this? Think about how we can make use of the thread co-participation count. On the left, we are connecting two students if they ever participated together; what about raising the required co-participation for a link to something higher? This will remove the weaker connections in the graph, allowing us to find the pairs who co-participated the most. On the right in illustration 8.13, you can see what happens with the link threshold set to 3. Now we can clearly identify communities of students who tended to participate together (at least in pairs). This is a very simple example illustrating the starting point of **community detection** in a social network.

It should make sense that the SLN in MOOCs follows this structure. The massive scale makes it hard for students to establish connections with a large percentage of their peers. Even among peers who have co-participated, the asynchronous form of communication makes it harder for them to maintain strong connections.

Helping Teachers Help Their Students

One of the ultimate goals in studying SLNs is to find ways for more effective teaching and learning. If given access to the types of visuals discussed here, what tasks may a teacher be able to accomplish more effectively?

For one, these graphs could help a teacher identify struggling students. Those who tend to ask many questions are likely not receiving their ideal learning experience from the content. Students in this category who also have a low in-degree in the (weighted, directed) student-to-student graph probably really want to learn the content but are not getting the help needed from their peers (e.g., student C in illustration 8.14). These individuals may stand to gain quite a lot from direct instructor intervention.

On the flip side, the teacher could also identify "student teachers" with these visuals. Those who answer many questions correctly are not only knowledgeable about the material, but are willing to help those who are not. A person with a high out-degree in the student-to-student graph, and high "importance"—which could be defined as it is through PageRank

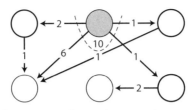

Illustration 8.14 A teacher can use the student-to-student graph to find strugglers and experts within the social learning network.

back in chapter 5—is likely in this category (e.g., the highlighted node in illustration 8.14). These students can be asked to help some of the struggling students.

These visuals could also help a teacher identify which course topics are receiving the most (and least) attention from students. The teacher may want to focus her own attention on those topics. The student-to-thread bipartite graph can help identify these, for example, by locating those threads with the highest degree.

In addition to showing visuals to teachers for them to draw conclusions from, is it possible for some discoveries to be automated, particularly those more readily identified by machine intelligence in the first place? There is active research, for example, on building algorithms to make predictions early in a course of what a student's eventual learning outcomes will be at his/her current rate. Some of these algorithms use collaborative filtering like we saw in chapter 7, except rather than predicting user ratings on movies, the objective is to predict student performance on quizzes, variation in student forum participation over time, or other outputs of interest.

A recent, interesting finding on this topic is that certain attributes of student behavior—in particular, specific sequences of actions that a student makes, and places that she visits while watching lecture videos—can predict the scores she will get on quizzes and exams. Recent research has shown, for example, that it is possible to obtain 70–80% accuracy in predicting whether a student will get a question right or wrong based solely on the behavior they exhibit while watching the video corresponding to the question. If a computer flags those students at the beginning of the course who are forecasted to perform poorly, teachers will know whom to help before it is too late. Considering the proportion of people who ultimately fail to complete a MOOC, each 1% improvement in accuracy translates to several dozens of students being correctly identified in advance.

Visualization, recommendation, and prediction are just three pieces of social learning networks. At the end of the day, remember that the underlying efficacy of an SLN hinges on the notion that a student's posting of a question will result in the generation of an accurate answer from her peers. The sustainability of an SLN is predicated on the hope that student needs can be partially satisfied from within, by the amalgamation of contribution and participation from the massive crowd of learners.

Summary of Part III

In this part of the book, we have explored three networked activities: rating products, recommending movies, and learning socially. This seemingly diverse set of applications each rely on some variation of the wisdom of crowds: as more opinions or knowledge about some truth are collected, we can make higher-quality estimates on what that truth is under the right circumstances.

In chapter 6, we saw that the wisdom of crowds is an effective tool in generating an accurate estimate of the underlying "ground truth" from a set of user ratings. When the individual sample size isn't large enough, techniques like Bayesian adjustment can be applied to weigh it against the larger population. These tools were introduced as we discussed some of the principles behind how Amazon ranks products, though their actual formula that goes from rating to ranking remains a secret.

We then turned to the recommendation problem in chapter 7. Here we saw how Netflix plays a "mind-reading" game to predict unknown user-movie ratings, using the past history of ratings Netflix has collected and stored in its database. We walked through the concepts behind how Netflix forms a baseline from the past history of the specific user and movie in question, and subsequently adjusts this based on similarities across different users or different movies.

Finally, in chapter 8 we explored social learning networks (SLNs), which are types of social networks that form among students collaborating on educational topics. We discussed massive open online courses (MOOCs) and current research looking to represent, analyze, and leverage the SLNs emerging in these scenarios. At the core of an SLN is the hope that the crowds of students can work together to solve one another's problems.

PART IV

CROWDS ARE NOT SO WISE

We just learned about the wisdom of crowds, which is the notion that aggregating information from a mass of people often results in better decisions than can be made by individuals. Inherent to this, though, is the assumption that everyone's opinions are independent. In what situations will they not be independent? What happens when they aren't?

In fact, in many scenarios, what other people think will influence your actions. Like that YouTube video you watched because everyone was talking about it, that iPad you bought because everyone else had one, or that answer you put for a homework problem when your peers insisted it was correct. Further, the more people you see that have followed through with such an action, the more the temptation increases for you to do the same.

In light of this behavior, our focus in this part of the book will be on the dependence of opinions in social networks. In chapter 9, we look at viralization, and in chapter 10, we talk about social influence. By the end of this part, you probably will be convinced that in certain situations, it can actually be relatively easy to influence crowds.

9

Viralizing Video Clips

In the space of user-generated video content, YouTube (see illustration 9.1) reigns as the dominant sharing site. Browsing through its selection, we can find videos ranging from short clips of amazing plays in sports, to the music of self-proclaimed YouTube artists, to lectures on educational topics. As of 2015, hundreds of millions of hours worth of videos were watched on YouTube every day.

One term you've probably heard used in the same sentence as YouTube is **viralization**. You may have wondered, what exactly does it take for a video clip to become viral? We'll look at that in this chapter. In doing so, we'll recognize how YouTube viewing is a good example of dependencies created by information spread.

YOUTUBE AND VIRALIZATION

Before jumping into a discussion on viralization, let's take a look at YouTube's evolution. The company was founded in February 2005 by three former PayPal employees. From its inception, the site grew at a rapid rate: by July 2006, it was getting 65,000 new videos and 100 million views daily. A few months later, in November 2006, the company was acquired by Google for $1.65 billion.

Within a few years, people were watching videos on YouTube so much that the site became a search engine second in size only to Google itself. The progression of YouTube's average daily video view count can be seen in illustration 9.2: in October 2009, it reached 1 billion, and by January 2012, the count had increased fourfold. This comes as no surprise if you've ever experienced the addictive nature of the recommendation sidebar on the site, bringing viewers a continuous stream of relevant short clips for up to hours at a time before they click out of it.

Illustration 9.1   YouTube's trademarked logo.

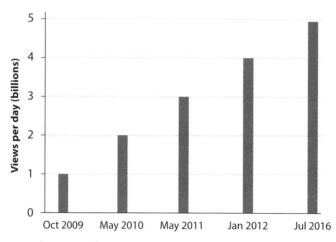

Illustration 9.2   There were four months between 2009 and 2012 when the average number of video views per day on YouTube hit another billion. As of July 2016, this had reached just under 5 billion views.

By mid-2016, more than 1 billion people were visiting YouTube each day and registering almost 5 billion video views. On average, 400 hours worth of new content was being uploaded to YouTube every single minute, which is almost 66 years of content per day (in other words, if you collected all the videos that will be uploaded in the next 24 hours, it would take you 66 years of your life to watch them all!). YouTube has become a viral phenomenon itself, like the video clips on its site aim to become.

Viral Style

What makes videos go viral? This question doesn't have a simple answer, but that hasn't prevented people from studying it.

We talked in part III of the book about how websites can capture and store the actions that users make on the site. YouTube is no exception:

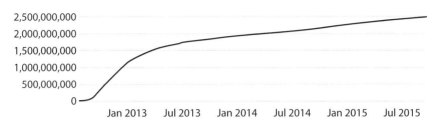

Illustration 9.3 Aggregate views over time for the music video "Gangnam Style" by PSY, on YouTube.

it can log user behavior, including interactions with its video player. This data can be analyzed to see how people watch videos and which videos have gone viral. Some of YouTube's analytic tools for highlighting overall viewing behavior, such as YouTube Insight, have been made available for public access too.

What is the most notorious video that achieved viral status? That would be "Gangnam Style," a 4-minute music video by the singer PSY. Released in July 2012, "Gangnam Style" became the first video ever to break 1 billion views, which it did in just 5 months (by December 2012). It proceeded to hit the 2 billion mark in less than 2 years (by May 2014). The graph of its views over time, provided by YouTube's analytics, can be seen in illustration 9.3.

Since 2013, 12 other YouTube videos have also broken the 1 billion barrier. Second to "Gangnam Style" in view count as of early 2016 was Taylor Swift's "Blank Space," with 1.3 billion views. But "Gangnam Style" was, at the time of this writing, still the only video to have hit 2 billion. In fact, when it surpassed 2,147,483,647 views in December 2014, YouTube appeared to have literally lost count, as the value displayed on the main page came to a grinding halt.

Why did YouTube stop counting at this seemingly random number? This value is the largest that can be stored using 32 bits, which is the amount that YouTube had reserved for the view counter of each video. Nobody ever anticipated that a single video would amass so many views that 32 bits would no longer be enough. YouTube quickly fixed this problem by upgrading to a 64-bit counter, giving them a new maximum of 9,223,372,036,854,775,808. So until PSY, Swift, or someone else reaches the quintillions, YouTube's counter should be safe.

Bringing Viewers to Videos

How does a video like "Gangnam Style" get so popular? We can start answering this question by looking at the four main paths that may lead a viewer to a particular YouTube clip in the first place:

- a *search* with terms the video is tagged with, on sites like Google,
- a *referral* (from, e.g., email, Facebook, or ads promoting the video),
- a *subscription* to the YouTube channel that posts the video, and
- a *recommendation* to the video given in YouTube's sidebar.

Subscription and recommendation often play a bigger role in determining a video's popularity than the number of likes and dislikes it has. Subscription is straightforward to understand, but how does YouTube generate its recommendations? Does it use a collaborative filtering algorithm like Netflix does for recommending movies? Or maybe a PageRank-style algorithm to rank the clips by "importance"?

It turns out that neither algorithm would translate well to this application. Unlike Netflix movies, YouTube videos are typically short in length and lifecycle and have variable viewing behavior, which would make it hard to establish a consistent system for users rating clips. For the PageRank approach, we would need to "link" clips together somehow (e.g., by searching a video's description for hyperlinks to other clips or by comparing tags between videos for matches in keywords). But tags and descriptions can be rather unreliable in quality.

YouTube video recommendation is believed to be different and also much simpler. Remember co-participation from chapter 8, when we weighted links between students by their co-participation in discussion threads, and vice versa? YouTube keeps track of the **co-visitation count** for pairs of videos, which is the number of times both videos were watched by a viewer in some recent time window (say, over the past 24 hours). So if 100 people have watched both videos A and B in the last day, we could link A and B with a weight of 100. Using this approach, a weighted video-to-video graph can be constructed. You can see an example in illustration 9.4.

YouTube seems to take this co-visitation graph and combine it with a match of the keywords in video titles, tags, and summaries to generate your recommendations. It has also been observed that often only those videos with a watch-count number similar to, or slightly higher than, that of the

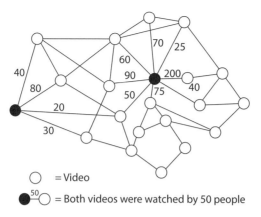

Illustration 9.4 YouTube recommendation leverages the co-visitation count between videos. Here, each node is a video, and a link's weight is the number of people who watched both videos over some duration.

current video are shown in the recommendation page. This makes it easier for widely watched videos to become even more widely watched: a positive feedback!

Defining Viral

What exactly is meant by "viral"? Although there is no commonly accepted definition, it usually implies that the video's total views over time look something like curve (c) in illustration 9.5. There are three important features here:

1. a high total view count,
2. a rapid increase of sufficient duration, and
3. (sometimes) a short time before the rapid increase begins.

There is no golden formula you can follow to guarantee your video will become viral. Still, models that have been developed for **information spread** can give interesting insight into why viralization may occur. These idealized models have been used to analyze the spread of "items"—ranging from physical products to diseases—through populations. Thinking of YouTube videos as items, we will soon look at one simple yet illuminating model for information spread in this chapter.

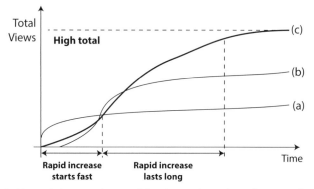

Illustration 9.5 Typical shapes observed for the total number of views of a video over time. Video (a) stays at a low level. Video (b) rises very quickly, but then flattens out rapidly too. Video (c) has a reasonably rapid increase of long duration, and stays that way for a while.

Popularity

Let's think about the factors that attract people to an item in the first place. One, of course, is the **intrinsic value** that the item brings to a person. Some may like it, regardless of what others think about it.

In many instances, though, a person's decision to obtain an item will depend on what other people have done. A **network effect**, as it's called, can occur for one of two reasons. First, the value of a service or product may depend on the number of people who use it. What use would a telephone be to you if nobody else had one? Would Facebook still be interesting if you were the only person who used it? These products and services have a **positive network effect**: as more people use them, they become more valuable to each individual.

Second, knowing what others think about an item can affect your decision. Have you ever watched a movie because your friends told you that it was good, irrespective of whether it is the genre you typically like? In these situations, peoples' opinions and decisions are influenced by others. The crowds are no longer "wise" as they were in part III, because the assumption of independence that we made there no longer holds. Resulting instead is the **fallacy of crowds**.

Which of these factors applies to a person choosing to watch a video on YouTube? The site itself does have a positive network effect, but that doesn't really have a bearing on whether someone will choose to watch a particular

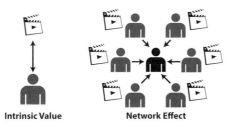

Illustration 9.6 Two factors affecting the popularity of a YouTube video are its intrinsic value and the network effect.

video. More influential are the other two components (see illustration 9.6): the intrinsic value of the clip to the person (i.e., whether it matches her preferences), and the fallacy of crowds (i.e., whether she sees a lot of other people watching it). The latter is based on a network effect that will spread the video viewing through the population and therefore has a larger impact on a video going viral.

Quantifying network effects is no easy task. They are dependent on the individual, the item, and the situation of interest. Next, we are going to look at a model for **information cascade**, which is a toy example of the fallacy of crowds.

THE FALLACY OF CROWDS: INFORMATION CASCADE

What would you do if you saw someone standing on a street corner looking up at the sky? You would probably think she had a nosebleed and go on with your business. But what if you saw ten people standing together looking up at the sky, as in illustration 9.7? Then you would probably stop and look, thinking that something may be wrong. This makes the crowd even bigger, so the next person passing by will see eleven people, which makes it even more compelling to stop and look.

This is a classic example of an information cascade, where people follow the actions of the crowd and ignore their own internal reasoning. Information cascades arise when the independence assumption about opinions (behind the wisdom of crowds in part III of the book) breaks down. In fact, it embodies the exact opposite case: instead of complete independence in decision making, decisions become completely dependent on what has happened before.

Illustration 9.7 Everyone is looking up at the sky on a street corner. This would make passerbys think there might be something wrong, influencing them to look, too.

One could probably write a whole book on all the examples of information cascades: from stock market bubbles, to fashion fads, to the collapse of totalitarian regimes throughout history. How do they apply to videos going viral? You are more likely to stumble on a video that is already popular. Even if it doesn't match your taste, you may be compelled to see what it's all about. You might decide to stop viewing it if you don't like it, but this will still count toward the viewing number shown next to the video and partially determines its place on the recommendation page. A higher view count will in turn influence more people, and this accumulation keeps on building.

Making Decisions in Sequence

We need to look at the process that eventually triggers an information cascade. In **sequential decision making**, each person gets a private signal (e.g., my nose starts bleeding) and releases a public action (e.g., tilt my head to the sky). Subsequent users can observe the public action but not the private signal. If we come to a time where there are enough public actions of the same type (e.g., ten people looking at the sky), then all later users will

Illustration 9.8 Positive feedback in sequential decision making. As the number of people displaying the same public action increases, it makes it more tempting for the next person to mimic that action, increasing the number performing the action, and so on.

ignore their own private signals and simply follow what others are doing. At this point, a cascade has been triggered.

How many public actions are enough for a cascade? That depends on the situation at hand. It's probably much harder to get everyone to watch your YouTube video than it is to get people to look up at the sky, for example. It also depends on the people involved. A cascade can start quicker if people are more ready to rely on others' public actions in their reasoning.

A cascade can accumulate to a large size through **positive feedback**. You can see this in illustration 9.8: more people displaying the same public action gives the next person more incentive to follow, which makes the group larger, thereby creating more incentive, and so on. Remember our discussions of negative feedback in part I of the book? Positive feedback is the opposite. In the former, we systematically counteract an effect to reach an equilibrium in the network (through, e.g., distributed power control in chapter 1 or usage-based pricing in chapter 3). In the latter, the effect feeds off its own unabated influence, generating more influence, and continues to grow larger. Both types of feedback are important themes in networking.

Will the public action that everyone follows in a cascade be right or wrong? It could be either. A wrong cascade (e.g., everyone is looking up,

Illustration 9.9 In this thought experiment, some people are lined up, and one at a time, they are called up to write their guess on the chalkboard. This is what the board might look like when the third person goes up to it.

but there's nothing of interest in the sky) is the epitome of the fallacy of crowds. But a cascade is also fragile: even if a few private signals are leaked to the public (e.g., one single person on the street corner shouts "I am looking up at the sky because I have a nosebleed!"), it can quickly disappear or even reverse direction. Why? Since people are following the crowd, they have little faith in what they are doing, even though many are doing the same thing.

Several models for sequential decision making have been proposed over the years. We'll take a look at a simple one next.

The "Number-Guessing" Thought Experiment

Consider a group of people lined up to play a game in which they will guess a number. The moderator has picked a number, either 0 or 1, to be the true one. One at a time, each person comes up to a blackboard, where she is to write down what she thinks the number is (illustration 9.9).

When a person comes up, the moderator will show him/her a card with either a "0" or a "1" written on it. This serves as the person's private signal. If the true number is "0", the moderator has a chance, say 80%, of choosing the "0" card, and 20% of choosing the "1" card. If it's "1", the moderator has an 80% chance of choosing the "1" card and a 20% chance of the "0" card. There's no guarantee that the number a person is shown will be right, but everyone is told that there's a higher chance that the card they are shown is right than wrong.

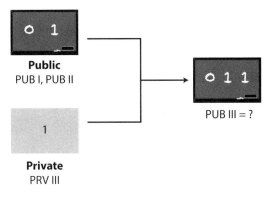

Illustration 9.10 The third person sees the public actions of the first two people (`PUB I` and `PUB II`), and receives her own private signal (`PRV III`). From this information, she writes down her guess on the chalkboard (`PUB III`).

Each person's guess on the blackboard is her public action. When a person goes to make a guess, she gets to see the public actions of everyone who went up before her, but she does not see the private signals they were shown. As an example, illustrations 9.9 and 9.10 show when the third person goes up to the board:

- She is shown a private signal on a notecard, which we'll call `PRV III`.
- She sees the public actions of the first two users on the blackboard, which we'll call `PUB I` and `PUB II`.
- Using this information, her task is to make her guess, `PUB III`, and write it on the board.

Let's say Alice is the first person to go up. What should she do? There's nothing currently on the blackboard, so all she has to go by is the number on her notecard. She knows this number is more likely to be right than wrong. So if she sees a 0, she will write 0, and if she sees 1, she will write 1.

Now Bob is the second to go up. How is his situation different from Alice's? Not only does he see both the public action that Alice wrote (`PUB I`) and his own private signal (`PRV II`), he also knows how Alice reasoned. He cannot see her private signal, but he knows it must be the same as `PUB I`, because Alice had no other information when she guessed. So Bob really knows two private signals, `PRV I` and `PRV II`:

- If they are both 0, then obviously he will write down 0. This is just a stronger case of what Alice was facing.

- If they are both 1, then similarly, he will write down 1.
- But what if they are different? Then he has no indication as to which number is more likely to be correct. It's his card against hers. In this case, he would flip a coin and randomly write down 0 or 1.

Now comes the first chance of an information cascade starting. When Cara—the third person—goes up to guess, what information does she have? She has one private signal, PRV III, and two public actions, PUB I and PUB II. Cara needs to compare PUB I and PUB II.

First, what if they are different? Then Cara knows Alice's and Bob's private signals must have been different, too. Bob must have seen a mismatch and guessed randomly. These two conflicting private signals cancel out, leaving Cara in exactly the same situation that Alice—the first person—was in. She'd then just guess based on her own private signal, PRV III.

Now, what if PUB I and PUB II are the same?

- If Cara's PRV III matches, then it's a no-brainer: she knows two signals say her number (hers and Alice's), and another could have matched (Bob's). So, she should pick this number for PUB III.
- Here's the really interesting part. Even if Cara's PRV III doesn't match PUB I and PUB II, it turns out that her best guess is to ignore her private signal and go with the last two public actions anyway!

So, if the first two people (Alice and Bob) write down the same guess, then an information cascade starts. The third person's (Cara's) rational choice is just to keep with the crowd. If the third person went with the crowd, then the fourth person will, and so on (until something else comes along to break up the cascade).

Why does a cascade start after the first two people? Let's break it down logically. Cara knows what Alice's private signal is. Being different from hers, they will cancel out, so Cara's decision comes down to what she can surmise about Bob's private signal. Going back to his decision, there's two ways in which his public action could have matched Alice's:

1. Bob's PRV II matched Alice's PUB I (this means Bob's public action is his private signal), or
2. Bob's PRV II didn't match Alice's PUB I, but when he chose randomly, he landed on PUB I (this means Bob's public action is not his private signal).

Illustration 9.11 When the third person and fourth person both guess 1, a cascade of 1s is triggered.

The first case is more likely. So, it is more likely than not that Bob guessed his private signal. Therefore, Cara's best bet is to guess whatever PUB II is, and we have an information cascade started.

What if no cascade has started after the first two people? Then everything restarts, and a cascade could just as well start after the next two people. And then the next two, and so on. All it takes is some even-numbered person to show the same public action as the odd-numbered person right before her. In illustration 9.11, a cascade of 1s is triggered after the third and fourth people show public actions of 1.

STARTING A CASCADE

How long can we expect it will take for a cascade to start? How easy can it be to break a cascade? We'll take a look at these points here, by going through some numerical examples of our number-guessing experiment. In you prefer, you can also skip the following calculations and go straight to the next section.

The First Pair of Guessers

Alice and Bob are the first pair of people to guess in our experiment. Let's say the moderator has decided on 1 as the correct number and that the chance that she shows each person a 1 as their private signal is 80% (which we'll call the moderator's chance).

The easiest way to break down the different types of cascades is through a tree diagram as in illustration 9.12, which shows the six different

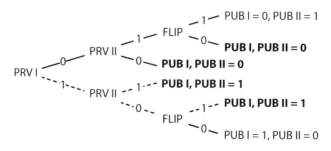

Illustration 9.12 All the possible scenarios arising from the first two guessers. The four highlighted will trigger information cascades.

possibilities based on Alice's and Bob's private signals. Two possibilities result in an incorrect cascade (both public actions 0), two result in a correct cascade (both public actions 1), and two result in no cascade (public actions different). For example, PRV I = 1, PRV II = 0, and FLIP = 1 traces out the case of Alice getting (and guessing) 1, followed by Bob getting 0, flipping a coin and guessing 1, which triggers a correct cascade of 1s.

What is the chance that no cascade will have been triggered at the end of their turns? For this to happen, we need the first two public actions to come out different, which is true if (i) PUB I = 0 and PUB II = 1 or if (ii) PUB I = 1 and PUB II = 0. From the diagram, you can see that these will happen for the following combinations of private signals:

- (i) PRV I = 0, PRV II = 1, FLIP = 1: What is the chance that Alice gets a private signal of 0? With an 80% chance of seeing a 1, there's a 20% chance of a 0. Then how about for Bob getting a 1? It's 80%. What is the chance the coin flip lands on 1? 50%. Because each of these three independent events must happen, we multiply them to get the total chance of this happening: $0.20 \times 0.80 \times 0.5 = 0.08$, or 8%.
- (ii) PRV I = 1, PRV II = 0, FLIP = 0: This is the reverse of the logic we just saw. What is the chance Alice sees a 1? It's 80%. How about Bob seeing 0? 20%. And the coin flip landing on 0? 50%. These chances give the same product: $0.20 \times 0.80 \times 0.5 = 0.08$, or 8%.

Since (i) and (ii) both lead to no cascade, we add them to find the total chance: $8\% + 8\% = 16\%$.

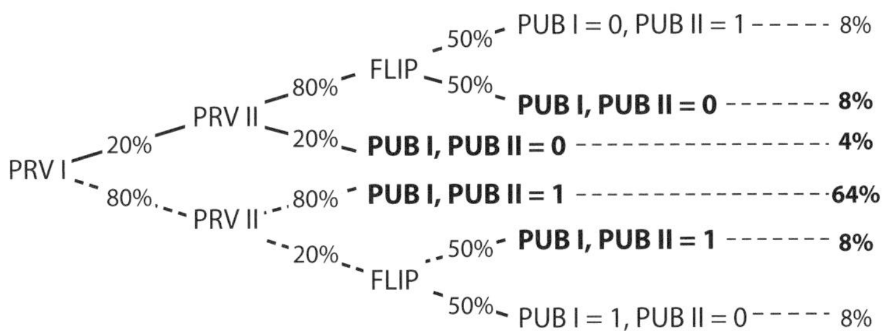

Illustration 9.13 The chance of each of the six outcomes happening in our example is shown to the right of the tree diagram.

Now, what is the probability that a cascade will occur? That's easy: 16% is the chance of it not occurring, so $100\% - 16\% = 84\%$ is the chance that it will. Can we break down this 84% further? Yes, because two different types of cascades could occur: a correct one (1s) or an incorrect one (0s). Remember that an incorrect cascade is the epitome of the fallacy of crowds in this model.

To find these chances, in illustration 9.13, the chances of the six different outcomes are broken down. For each case, we just multiply down the branch: for example, PRV I being 0 has a 20% chance, PRV II being 1 has an 80% chance, and FLIP being 0 has a 50% chance, which gives $0.2 \times 0.8 \times 0.5 = 0.08$ or an 8% chance of this sequence happening. Adding up the relevant possibilities, the chance of a correct cascade is $64\% + 8\% = 72\%$, and that of an incorrect cascade is $4\% + 8\% = 12\%$.

The chance of a correct cascade, then, is pretty high. How come? Remember the assumption we made about the moderator at the beginning? There's an 80% chance that she shows the correct private signal (1) to each person. That's pretty high too. If we lowered it, as it turns out, both the incorrect cascade and no cascade would become more likely. For more on this relationship, and a more detailed calculation, check out Q9.1 and Q9.2 on the book's website.

Future Pairs of Guessers

After Alice and Bob, the chance of no cascade is 16%. How about after Cara, then? Remember that the third person cannot herself trigger a cascade; if none was triggered after Bob, then Cara starts from scratch with no

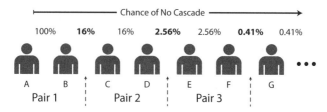

Illustration 9.14   In the number-guessing experiment, as more pairs of people have guessed, the chance we won't wind up in a cascade gets lower.

information, effectively in Alice's shoes. So after the first three people, the probability of no cascade is still 16%.

How about after Dana? Now we have two ways that a cascade could be triggered: first after Alice and Bob, and then after Cara and Dana. To have no cascade at the end, we need the first pair (Alice and Bob) and the second pair (Cara and Dana) to not trigger one. So we multiply the chance of each not causing one: $0.16 \times 0.16 = 0.0256$, or 2.56%. After Dana, then, there's more than a 97% chance that we will be in a cascade.

Then how about after Evan? As the first person in the third pair, he cannot cause a cascade, so the chance stays the same: 2.56%. And after Frank? Now we have three chances of a cascade: after Bob, Dana, and Frank. So we multiply three times: $0.16 \times 0.16 \times 0.16 = 0.0041$, or 0.41%. That's a very small chance of not having a cascade.

You probably see the pattern by now, given in illustration 9.14. To get the chance of still having no cascade after "N" pairs, we multiply 0.16 "N" times. For five pairs, it becomes $0.16 \times 0.16 \times 0.16 \times 0.16 \times 0.16 = 0.000105$ (i.e., about 1/100 of a percent). For 50 pairs, the decimal has 37 zeros before the first significant digit!

Clearly, this chance is going to zero, and very quickly at that. So after a few pairs, we are more or less guaranteed that we will have a cascade. At this point, the decisions of future people have become entirely dependent on what is on the board already.

But remember, a cascade can either be correct or incorrect. Can we calculate how likely each type is after a given number of pairs have guessed? We could, but those formulas require a bit more math than just multiplying repeatedly. As it turns out, the number of pairs that have guessed does not change which type of cascade is more probable all that much. With the moderator's chance at 80%, for example, the chance of an incorrect cascade goes from 12% after one pair to roughly 15% after three pairs, while the chance of a correct cascade goes from 72% to roughly 85%. A correct cascade is still much more likely.

The moderator's chance itself is what affects the type of cascade, irrespective of the number of pairs. If this dropped to 60%, for example, then the chance of having an incorrect cascade eventually would be more than 35%, with that of a correct cascade being less than 65%. At a 50% probability, they will both be equally likely!

For a plot of these relationships, check out Q9.3 on the book's website. Suffice it to say that if our goal is to trigger a correct cascade, we have to hope the moderator would show the right value a lot more often than the wrong one. This point is somewhat counterintuitive, because when the number of pairs is large, we expect a large amount of information to be available (i.e., more public actions), and that a correct cascade should be very likely to happen. But cascades block the aggregation of independent information, which was so important to our discussion around the wisdom of crowds in part III. All it takes is one pair that displays two 0s or two 1s to render all future public actions meaningless.

THE EMPEROR'S NEW CLOTHES

You now see how easy it can be to start a cascade. Once triggered, how long will one last? Forever, unless there is some kind of disturbance (e.g., a release of private signals). How many disturbances would it take? Interestingly, even a few will often suffice, no matter how long the cascade has been going on. Despite the number of people involved, everyone knows that they are just basically playing a game of follow the leader to maximize their chances of guessing correctly.

The **Emperor's New Clothes effect** encapsulates the fragility of an information cascade. This name arises from the nineteenth-century short story by Hans Christian Andersen (illustration 9.15) in which a vain emperor is told that his new "clothing" is of the finest fabric, invisible only to those who are unfit for their positions. In reality, there are no clothes at all. While everyone plays along (i.e., their public actions), not wanting to seem unfit (i.e., their private signals), it only takes one kid's shouting "hey, he is wearing nothing at all!" before everyone becomes more confident that the emperor is truly disrobed in public.

Illustration 9.15 The Emperor's New Clothes effect is the notion that information cascades can be broken relatively easily.

Going back to our number-guessing experiment, how do we break a cascade? Suppose a cascade of 1s has started after the first pair. Some time later, it's Frank's turn to guess, and he gets a 0 as his private signal. As his public action, he guesses 1, but he also shouts out that his signal was 0.

Now, Greg is up, and he also gets a private signal of 0. He has the following information about private signals:

- On the one hand, there's at least one 1, from Alice. But he cannot be sure if Bob had a 1, because Bob could have gotten a 0 and flipped.
- On the other hand, there's at least two 0s: his and Frank's.

So, what will Greg do? He'll guess 0, because there's more evidence of this being correct. This guess breaks the cascade. Remember, a cascade represents only what happened with a few people right around the time it started. If everyone knows that, another block of a few people may be able to break it.

There Are Many Additional Factors

In the number-guessing experiment, it only takes two people to start a cascade. More generally, the size of the crowd needed to cause a person to ignore their instinct is dependent both on (i) the scenario and (ii) the individual. For example, to follow the crowd, a person would probably need to see more heads tilted as a passerby on a street corner than matching numbers on a blackboard. And a passerby who is in a hurry may take even more convincing than one who is bored and curious.

But perhaps the most far-reaching assumption we have made here is that everyone acts rationally. We have assumed that everyone can, and will, decide what the best guess is depending on the information they have. Will this assumption always be true? Definitely not. What each person should do can be quite different from what she does in reality. Researchers have observed that the number-guessing thought experiment does not work as the theory predicts, probably because most humans do not go through all of this reasoning about probabilities in their heads.

How can we translate sequential decision making to viralizing a YouTube video? Not easy, but the main idea should be clear: you want your video to undergo an information cascade, so that when a person sees or hears about it (i.e., the public action), they will most likely watch it, irrespective of whether it matches her intrinsic interest (i.e., her private signal). How many public actions does it take before a person will watch your video automatically? Does such a number even exist? Even if it does, it will be different for everyone, depending on how malleable the person is. These are all interesting questions that don't yet have clear answers.

10

Influencing People

What was the main message from chapter 9? Exposure to public opinion breaks the independence needed for the wisdom of crowds. In the ensuing fallacy, people are influenced by the actions of others.

In this chapter, we continue with our theme of influence, this time paying more attention to the underlying graph of social networks. Our study on viralization was for population-based scenarios, where it's assumed that someone's public action has the same effect on the next person regardless of their relationship (as shown in illustration 10.1). Turning to social networking sites like Facebook and Twitter, our discussion will center around topology-dependent, rather than population-based, aspects of influence.

SITES OF SOCIAL INFLUENCE

Friendship on Facebook

What is the largest social networking application today? That would be Facebook (illustration 10.2). The 1.65 billion people who accessed Facebook at least once in March 2016 represent more than one-fifth of all the people on earth.

In January 2004, Mark Zuckerberg launched a social networking site called "Thefacebook" for his classmates at Harvard University. Having quickly attracted more than half of the undergraduate population there, within 3 months it started expanding to other Ivy League schools and gradually gained the attention of most universities in the United States and Canada. In 2005, Facebook dropped "the" from its name, opened up to high school students and soon after to company employees. It wasn't until September 2006, though, that the site began allowing anyone 13 years or older with a valid email address to join, which is the way it remains today.

In illustration 10.3, you can see the rapid growth in the number of monthly active Facebook users each year from 2004 to 2015. In 2012, the

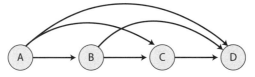

Illustration 10.1 Population-based relationship between people that we assumed in the number-guessing thought experiment in chapter 9. A link from A to B means that B obtains information (i.e., public action) from A.

Illustration 10.2 Trademarked logos of Facebook (left) and Twitter (right).

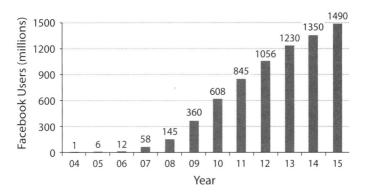

Illustration 10.3 Number of active Facebook users by year since 2004.

same year it broke 1 billion users for the first time, the company had its initial public offering, or IPO.

By 2015/2016, more than 4 million posts were being liked, 32 million items shared, and 240,000 photos uploaded to Facebook every minute. Accounting for 30% of all users, the largest age group using the site was 25–34 years old, who would have been high school- and college-aged when Facebook emerged in 2006. More users are from the United States than from any other country, but Facebook has gained some international foothold, too, with about 20% of its users from Europe and another 20% from Asia.

Facebook's "friend" feature is the function that made it so popular in the first place. Just think about all the connections that have been established among people through that "add friend" button over the years. These links, and the interactions they enable, bridge the gap between Facebook as a social media and a social networking site.

With Facebook users as nodes, we could create a graph by drawing links between all pairs of people who are friends. In 2015/2016, there were about 1.5 billion users, and the average person had about 350 friends. That means the graph would have 1.5 billion nodes, and $1.5 \times 350/2 \approx 260$ billion links (dividing by two because we don't want to count links twice). This would be an extremely large and complex structure, begging the question of what really constitutes a link between two people on Facebook. Friendship is certainly a concrete way of defining a link, but maybe some stronger notion is more meaningful, like the number of times the "friends" communicate?

Following on Twitter

Twitter (see illustration 10.2) is another popular social networking site, giving users the abilities to microblog—sending and receiving text-based "tweets" of up to 140 characters in length—and to "follow" other people. By the first quarter of 2016, about 1.3 billion Twitter accounts had been created, with 310 million qualifying as monthly active users.

First introduced in March 2006 by Jack Dorsey at Odeo, Twitter launched in July of that year and established itself as an independent company the following April. Like Facebook, the company experienced rapid growth from its inception, moving from roughly 400,000 tweets per quarter in 2007 to 100 million per quarter in 2008. By its ninth anniversary in 2015, it was handling almost 500 million tweets per day (that's almost 6,000 per second!).

Twitter tends to spike in usage during major events. During the East-Coast earthquake of summer 2011, tweets traveled faster than the earthquake itself from Virginia to New York. Check out illustration 10.4 for the level of Twitter activity just 10 and 80 seconds after this earthquake hit. As of 2016, Super Bowl XLIX played in February 2015 was the most tweeted event ever, with 25.1 million total tweets during the game.

As of May 2016, Katy Perry was the most followed person on Twitter with 88 million followers, Justin Bieber coming in second at 81 million.

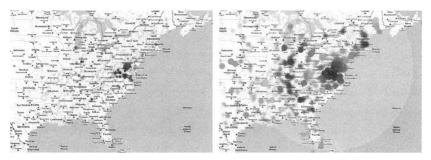

Illustration 10.4 Twitter activity 10 seconds (left) and 80 seconds (right) after an earthquake hit the East Coast of the United States in summer 2011.

For famous individuals, the one-way nature of following relationships on Twitter (i.e., you can follow them without them having to follow you) is particularly useful for broadcasting updates to fans. The "like" feature for celebrity pages on Facebook works like this, too.

Who Is "Important"?

Facebook and Twitter are just two of the many online social networking sites that exist today. Others have emerged around specific applications, from photo and video sharing on Instagram to business reviewing on Yelp to social learning in MOOCs (discussed in chapter 8).

A lot of studies have been done about how people behave and interact in these networks. Attempts have been made to answer the following two important questions:

1. How can we measure the influential power of individuals?
2. How can we leverage the knowledge of who's influential and who's not to actually influence people online?

Neither of these is easy to answer. There is a significant gap between theory and practice, one of the biggest that can be found in this book. But that hasn't stopped people from trying.

Regarding the first question, for example, some companies are charting the influential power of individuals on Twitter. How do they quantify a person's influence? There are several possibilities, for example, the number of followers, number of retweets, or number of repostings the person has. Some companies are also attempting to put together the entire social graph of Facebook, which would allow them to figure out which people in the network are the most "important." We look at ways of defining importance in this chapter.

As for the second question, how would a marketing firm, for example, use knowledge of influential power to help sell a product? They will try to get the product in the hands of the people who are believed to have the most social influence. They may incentivize a few influential individuals, or a large number of randomly chosen but reasonably influential people, hoping that when others see them with the product they will decide to buy it. Later in this chapter, we will see how finding the best people with whom to seed the product can be difficult and rather counterintuitive.

Beyond marketing campaigns, we could write a whole book on the utility of identifying and leveraging influential individuals. For an interesting historical anecdote, consider the so-called midnight rides of Paul Revere and William Dawes on the brink of the American Revolution in 1775. On the night of April 18, starting near Boston and Cambridge, both of them were to alert Americans that the British were planning an assault on Lexington and Concord. They each took different paths to Lexington, before meeting there and riding to Concord. Revere was able to alert powerful militia leaders on his route to Lexington, and as a result, he was more effective than Dawes in spreading the word. This helped the Americans win the first battle of the war the following day.

COMMON DEFINITIONS OF SOCIAL IMPORTANCE

How can we measure a person's importance in a social network? We start with the social graph of the network, where each node is a person, as in illustration 10.5. There are many possibilities for what the links in a social graph can symbolize, as we saw for social learning networks in chapter 8: they can be directed or undirected, weighted or unweighted, depending on what we want the graph to represent.

In illustration 10.5, the links are undirected (without arrows). Let's say that a link here means the two connected nodes "know" each other (e.g., Dana knows Cara, Evan, and Frank, while Ben only knows Anna). Facebook's graph of friends is also undirected, because if you're my friend, then I must be your friend, too. So were the co-participation graphs among students that we saw in chapter 8. In contrast, Twitter's graph of following relationships is directed: even if you follow your favorite musician, chances

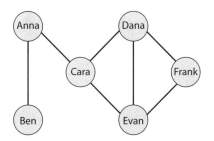

Illustration 10.5 Example of an undirected social graph that we use for computing centrality measures.

are he or she does not follow you. This was true of the webgraphs in chapter 5 as well.

With the graph in hand, how do we measure each node's importance? In fact, many different metrics have been proposed. Here, we look at three common measures of a node's **centrality**.

The Easy Way: Degree Centrality

The most obvious measure is **degree centrality**, or the number of nodes connected to the node in question. Back in chapter 5, we counted both the in-degree and out-degree for directed webgraphs. For undirected graphs, there's no distinction, so the task is even simpler.

What are the degrees in illustration 10.5? Anna's is 2: she is friends with Ben and Cara. Ben's is 1: he is only friends with Anna. Those of Cara, Dana, Evan, and Frank are 3, 3, 3, and 2, respectively. By this measure, then, we have Cara, Dana, and Evan tied as the most important, followed by Anna and Frank tied as the second most important, and finally Ben as the least important node.

Is this ranking reasonable? We can agree that Ben should probably be the least important: his only connection is to Anna. We can also agree that Dana and Evan should probably be equal in their importance, because they connect to the same people.

What about Cara, though? She provides the only connection that Anna and Ben have with the rest of the graph. Without her, the network would **partition** into two groups, as you can see in illustration 10.6. Additionally, without Anna, Ben would not be connected to the graph. If we are

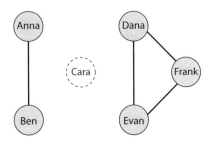

Illustration 10.6 If we remove Cara, then Anna and Ben will not be connected to the rest of the nodes anymore. This should say something about the centrality of Cara.

discussing centrality, then, shouldn't we give Anna and Cara more points since they are more vital to the graph's connectivity?

Another point that degree centrality fails to take into account is one we saw when ranking webpages in chapter 5: if a node is connected to many important nodes, it should probably be more important than if it is connected to many unimportant nodes. The PageRank algorithm (in chapter 5) solves these problems by having each node spread its importance to its neighbors. Can we apply that method here too? Indeed we could, and it would be a great exercise for you to do on your own. We will instead explore other approaches for defining centrality in a social graph.

Counting Paths: Closeness Centrality

Our second importance measure, **closeness centrality**, looks at how far a node is from its neighbors. To find it, we'll have to consider the distance between two nodes in a graph, which is the number of links in a **shortest path** containing them. With a **path** being a sequence of links connecting the nodes, the shortest path is one that uses the fewest possible links. We usually refer to a path by the nodes visited in traversing it, from start to end. So in illustration 10.5, the path from Ben to Frank that goes through Anna, Cara, Evan, and then Dana is (B, A, C, E, D, F). The length of this path is 5, since it has five links (finding shortest paths algorithmically is something we will explore in chapters 12 and 14).

There are more ways to get from Ben to Frank than by this path. For instance, we could have visited Dana before Evan: (B, A, C, D, E, F). We could have also gotten there using fewer links, since there's no need to use both Dana and Evan: (B, A, C, D, F) and (B, A, C, E, F) each have only four links.

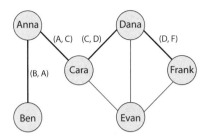

Illustration 10.7 One of the shortest paths between Ben and Frank is (B, A, C, D, F). The length is 4.

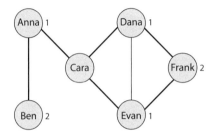

Illustration 10.8 The links and the lengths of the shortest paths for Cara.

These are the shortest paths between Ben and Frank, one of which is highlighted in illustration 10.7.

The closeness centrality of a node is based on the average of the shortest path lengths to each of the other nodes. Nodes for which this is small tend to be closer to other nodes, and are more important according to this metric.

Let's calculate this for Cara. First, we need to find the shortest paths between her and other nodes in illustration 10.5:

- What's the shortest path from Cara to Ben? There's only one path, (C, A, B). This has length 2.
- How about to Anna? The shortest path consists of one link, (C, A). This has length 1.
- To Dana? The shortest path is link (C, D), which has length 1.
- To Evan? Link (C, E), which has length 1.
- To Frank? The shortest paths are (C, D, F) and (C, E, F), with length 2.

You can see a summary of these distances in illustration 10.8. Now, what is the average? We get that by adding the distances together and dividing by the total number (5):

$$\frac{2 + 1 + 1 + 1 + 2}{5} = \frac{7}{5}$$

We are just about there, except for one minor detail. We want a smaller average (not a larger one) to give a higher centrality, to indicate that the person is closer. So we take the reciprocal:

$$\frac{1}{7/5} = \frac{5}{7} = 0.714$$

You can find the closeness centralities of the others in the graph using the same procedure. If you are interested in a walkthrough of the case for Dana, check out Q10.1 on the book's website. Here's a summary of the steps and the results:

$$\text{Anna}: \quad \frac{5}{1 + 1 + 2 + 2 + 3} = \frac{5}{9} = 0.556$$

$$\text{Ben}: \quad \frac{5}{1 + 2 + 3 + 3 + 4} = \frac{5}{13} = 0.385$$

$$\text{Dana}: \quad \frac{5}{2 + 3 + 1 + 1 + 1} = \frac{5}{8} = 0.625$$

$$\text{Evan}: \quad \frac{5}{2 + 3 + 1 + 1 + 1} = \frac{5}{8} = 0.625$$

$$\text{Frank}: \quad \frac{5}{3 + 4 + 2 + 1 + 1} = \frac{5}{11} = 0.455$$

By closeness centrality, then, Cara is the highest, followed by a tie between Dana and Evan, then Anna, then Frank, and finally Ben. Compared to degree centrality, a lot of the ties have been broken: Cara is first, and Anna now beats Frank. The promotion of Cara's and Anna's scores is justified by the fact that they are both vital to the graph's connectivity.

Closeness centrality is quite intuitive: the more people that are close to you, the more central you are to the network. Is it the best we can do? Some would argue that the ranking in our example still has some inconsistencies. First, why should Anna be less important than Dana and Evan? Dana and Evan really don't hold the graph together like Anna does (see illustration 10.9). Second, why is Cara's closeness only slightly higher than Dana's and Evan's? She is much more vital to the graph's connectivity.

There's another notion of centrality that's just as useful, and perhaps can fit this intuition better than closeness does.

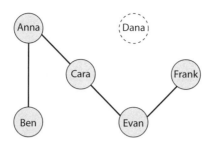

Illustration 10.9 We can remove Dana or Evan without causing the rest of the nodes to disconnect.

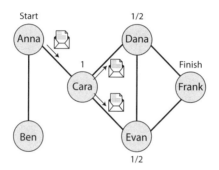

Illustration 10.10 If Anna wants to send a message to Frank, she must send it to Cara, while Cara can pass it to Dana or Evan.

Measuring Connectedness: Betweenness Centrality

Suppose that Anna needs to get a message to Frank over the shortest path possible through the network. Referring to illustration 10.10, there are two routes she could take:

- tell Cara, who tells Dana, who tells Frank, or
- tell Cara, who tells Evan, who tells Frank.

In either case, she needs to tell Cara. But then Cara has two possibilities: she can choose to tell either Dana or Evan. If we were going to assign importance points with respect to Anna's message to Frank, how would we do it? Cara may get the most, since both paths involve her. Only one of the two paths involves Dana or Evan, so they may each get half of what Cara is awarded.

As another example, what if Ben wants to send a message to Dana? Here, there's only one shortest path: send to Anna, who forwards to Cara, who passes to Dana. In awarding points, then, Anna and Cara may get the same amount. How many should they get relative to what we gave Cara, Dana, and Evan in the first example? Previously two paths were possible, but now there's only one, making it more critical. Without Anna or Cara, Ben couldn't message Dana; likewise, without Cara, Anna couldn't message Frank. So Anna and Cara may get the same amount of points for this one path as Cara did for both paths before.

Based on this, **betweenness centrality** considers a node to be more important when it lies on more critical paths between other nodes in the network. The more shortest paths there are from A to B, the less each will count toward another node's centrality, since each is less critical to the pair. If three shortest paths run from A to B, and two of these contain C, then what would C be awarded for this pair? 2/3. In our example, for the pair Anna and Frank, we would award Cara 2/2 (she lies on both paths), and Dana and Evan each 1/2 (they lie on one of the two paths). As for the pair Ben and Dana, Anna and Cara would each get 1/1 (they both lie on the only path).

Let's walk through computing the betweenness centrality for Cara. Before we start, intuitively, do we think hers will be high or low relative to the others? Probably high: she holds the two sides of the graph together. You probably know someone like that in your social network!

To find Cara's centrality, we need to consider each pair of nodes in the graph, except those with Cara. In doing so, we ask two questions: How many shortest paths are there between the pair? How many of these shortest paths contain Cara?

- Anna and Ben: How many shortest paths are there? Just one, link (A, B). Does it contain Cara? No, so she gets $0/1 = 0$ points.
- Anna and Dana: How many shortest paths? One, (A, C, D). Does it contain Cara? Yes. She gets $1/1 = 1$ point.
- Anna and Evan: Again, there's one shortest path, (A, C, E). Since it contains Cara, she again gets $1/1 = 1$ point.
- Anna and Frank: How many shortest paths? Two, (A, C, D, F) and (A, C, E, F). How many with Cara? Both, so she gets $2/2 = 1$.
- Ben and Dana: There's one shortest path, (B, A, C, D). And it contains Cara, so she gets $1/1 = 1$ point.

- Ben and Evan: Again, there's only one shortest path, (B, A, C, E), and it contains Cara, so she gets $1/1 = 1$ point.
- Ben and Frank: How many paths? Two, (B, A, C, D, F) and (B, A, C, E, F). Since both contain Cara, she gets $2/2 = 1$ point.
- Dana and Evan, Dana and Frank, Evan and Frank: Each of these pairs has one shortest path, (D, E), (D, F), and (E, F). None of them contain Cara, so she gets $0/1 = 0$ points for each.

These numbers are added up to get Cara's betweenness centrality:

$$0 + 1 + 1 + 1 + 1 + 1 + 1 + 0 + 0 + 0 = 6$$

You can go through this same procedure for the other nodes, too. If you're interested in the case for Dana, check out Q10.2 on the book's website. It turns out that Anna's betweenness centrality is 4 (second to Cara), Evan's and Frank's are 1.5 (tied for third), and Frank's and Ben's are 0 (tied for fourth; they don't lie on any shortest paths). Now, Cara is by far the most important: her value is 1.5 times higher than Anna's, and 4 times larger than Dana's and Evan's. Additionally, Anna is now more important than Dana and Evan: unlike the other measures, betweenness takes into account Anna's contribution to the graph's connectivity.

In illustration 10.11, you can see a summary of the different centrality measures we looked at for this example: degree, closeness, and betweenness. And we could throw in the PageRank importance score, too. Which one to use depends on the goal of using a centrality measure. The bottom line is that degree centrality is really rather naive, whereas closeness and betweenness will lead to rankings that better match our intuition about, and potential use of, who is important.

INFLUENCING SOCIALLY THROUGH CONTAGION

Keeping the notion of centrality in mind, let's continue our discussion on influence models from chapter 9, this time taking social graphs into consideration. We look at how a person's social relationships can influence her to adopt a product or item.

Consider the network in illustration 10.12, where there are eight nodes connected to one in the middle. Each of the neighboring nodes is in one

Node	Degree		Closeness		Betweenness	
	Value	Rank	Value	Rank	Value	Rank
Anna	2	2nd	0.39	5th	4	2nd
Ben	1	3rd	0.56	3rd	0	4th
Cara	3	1st	0.71	1st	6	1st
Dana	3	1st	0.63	2nd	1.5	3rd
Evan	3	1st	0.63	2nd	1.5	3rd
Frank	2	2nd	0.46	4th	0	4th

Illustration 10.11 Summary of the different centrality measures used in this example: degree, closeness, and betweenness.

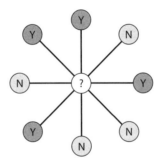

Illustration 10.12 Four of the center node's connections have switched (``Y''), and four of them have not (``N'').

of two states regarding whether they have adopted some product, service, or item: "Y" means yes, and "N" means no. Do the four nodes in the "Y" state present enough social influence to cause the center node to flip (i.e., to adopt the item) as well?

To put this in perspective, suppose you are the person in the middle, and the neighboring nodes are your close friends, some of whom have recently purchased the newest iPhone ("Y"), and others of whom have not ("N"). You can imagine that the more people you see who have it (assuming they are satisfied with it), the more influence that will have over you, and the more likely you will want to get one too. Is there a way for us to tell whether you will follow suit in the end?

A typical model is to set a **flipping threshold** for each node. This is the fraction of the node's neighbors that must have already flipped before that

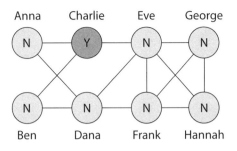

Illustration 10.13 Eight-node social graph used to illustrate contagion. You can think of this as a group of people purchasing a product. Initially, Charlie has the product and has ``flipped,'' while everyone else has not.

node will as well. In illustration 10.12, 50% (4/8) of the center's neighbors have flipped. So if her threshold is less than 50%, she would get the item too, but if it is something higher, then she wouldn't, because there is not enough social influence. For instance, if her threshold is 80%, she will need at least seven of her friends to have adopted it (since 7/8 > 0.8, but 6/8 is not).

Realistically, the flipping threshold is hard to estimate. It depends on a lot of different factors, as did the size of the crowd needed to start an information cascade in chapter 9. One factor is the item itself: less expensive and more attractive products would tend to lower the threshold, for example. Another factor is the individual: Bob could be relatively easy to sway, following suit as soon as one or two of his friends have, whereas Alice may never budge. There are also several network factors, like the strength of social ties between people, and what a link represents. For our purposes, we assume that we know the flipping threshold, and that it's the same for each node in the graph.

Let's consider the social graph of eight people in illustration 10.13. Charlie has flipped, while everyone else has not. Assuming a threshold of 50%, how will the graph change over time? This process is known as **contagion** and plays out as follows, in another idealized model.

First Step

In each time step, we go through person by person to see whether their flipping thresholds have been met, and if so, we switch them to "Y."

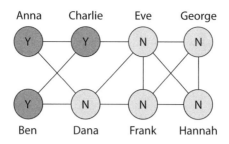

Illustration 10.14 Contagion after the first step.

- Anna: Initially, one of her neighbors is flipped (Charlie), and the other is not (Dana). Will she flip? Since exactly 50% have, Anna has been presented with just enough social influence, so she will.
- Ben: How many of his neighbors have flipped? One has (Charlie), and the other has not (Dana); 50% is enough, so he will flip.
- Dana: How about Dana? This one is a little tricky, because we just determined that two of her neighbors will flip. But, that doesn't happen until this round: at this point, none of her neighbors have flipped. With 0% influence, she will not either.
- Eve: One of her neighbors has flipped (Charlie), and the other four have not. How much influence does she see then? 1/5, or 20%. Since this is less than 50%, it is not enough.
- Frank, George, and Hannah: None of their neighbors have flipped.

The resulting graph is shown in illustration 10.14.

Second Step

Now what happens? The next step uses the updated graph (illustration 10.14):

- Dana: Two of her neighbors (Anna and Ben) have flipped, and two have not (Eve and Frank); 50% is just enough influence for her to flip.
- Eve, Frank, George, and Hannah: The situation hasn't changed.

In illustration 10.15, you can see the state of each node at the end of the second iteration. The left half of the network has flipped, and the right half has not. Are we on track to have everyone flip eventually? Let's see.

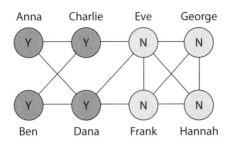

Illustration 10.15 Contagion after the second step.

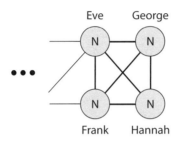

Illustration 10.16 The cluster of Eve, Frank, George, and Hannah is too dense to be penetrated by the outside social force. Eve has 60% of her links inside the cluster, Frank 75%, and George and Hannah 100%.

Third Step

What happens now?

- Eve: Two of her neighbors have flipped, and the other three haven't, making 40%. Not quite enough.
- Frank: One of his neighbors has flipped, and the other three haven't. This makes 25%. Again, not enough influence.
- George and Hannah: Still, none of their neighbors have flipped.

So, none of the nodes switch state in the third step. Why is this? Because each of these nodes has too many neighbors who have not flipped and not enough who have. Out of Eve, Frank, George, and Hannah, Eve has the highest fraction of links to those in the "Y" state, and the influence on her is only 40%. In other words, these four nodes form a social **cluster** that cannot be penetrated from outside (see illustration 10.16).

More generally, any group of nodes that has connections among themselves can be called a cluster. Taking a cluster of unflipped nodes, if each of them has too many neighbors inside, then it will be impossible to influence it. If you're interested in how to determine a cluster's **density**, check out Q10.3 on the book's website.

STRATEGIC MARKETING: MAXIMIZE COLLECTIVE INFLUENCE

Can you see the connection between contagion and strategic marketing? The objective is to maximize the amount of people who will buy our product by choosing the right people to seed it with in the first place. In illustration 10.15, we've reached half of the people.

What measures can we take generally to get more people to flip? If we know the social graph and trust the contagion model, there are a few possibilities. One would be to try lowering the flipping threshold: if it drops, even for just some nodes, we would have a better chance of influencing everyone to adopt at some point. Another option is to try breaking clusters. If we somehow cut social ties from within, we could more readily penetrate them from outside. Or we could try to seed nodes inside the clusters, to introduce social influence from inside.

Which of these options are doable? Moving the thresholds would require us to change peoples' preferences, while breaking links would call for changing social relationships. These are factors that a marketing company probably doesn't have control over. An alternative is to see whether we can pay nodes in clusters to adopt the product. Going back to illustration 10.15, for example, seeding any one of Eve, George, Frank, or Hannah would do the job.

More generally, suppose each person can be influenced with a certain amount of money. Presumably, those nodes that perceive themselves as more influential will require more money (e.g., a designer company might pay a famous celebrity a large sum of money to wear its clothing brand). Under a total budget constraint, which nodes should we seed to maximize the extent of flipping at equilibrium, while perhaps minimizing the time it takes to reach this equilibrium?

This is yet another hard question to answer. It would be great if we could find a way to trigger positive feedback as occurred for sequential decision making in chapter 9, where our seed set influences nodes to flip, creating enough influence to cause even more to flip, creating even more influence,

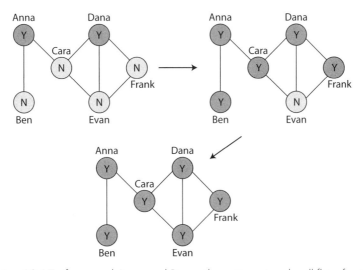

Illustration 10.17 If we seed Anna and Dana, the entire network will flip after two time steps.

and so on. If we could seed just one node, maybe we should go with the most important one based on one of the centrality measures? This is not always a safe bet. When we can seed more than one node, it becomes even more complicated. In general, we don't want to choose the most central nodes; rather, our objective is to choose those whose "combined" influential power is the largest.

Take our original social graph of six nodes in illustration 10.5. Cara was the most important both in terms of closeness and betweenness centrality. If we choose to seed Cara initially, what will the result be? After two time steps, we would have three nodes flipped: Anna, Ben, and Cara. But can we do better than that? Yes: Interestingly, seeding either Dana or Evan will cause all the nodes to flip after five steps! Those centrality measures are not always in line with maximizing influence, which makes it very difficult to come up with a robust solution for the choice of seed nodes.

What if we could choose two nodes to seed? We might figure, let's just take the two most important by betweenness centrality: Cara and Anna. What would happen? Actually not much more than when we just seeded Cara. So, does there exist a set of two nodes that will cause the whole network to flip faster than if we just seeded Dana or Evan? Yes: consider what happens when we seed Anna and Dana, as in illustration 10.17:

- After one step, we cause everyone to flip except Evan.
- After two steps, Evan also flips, so we have achieved our goal.

Again, choosing based on centrality is not always going to be the best strategy. The important thing is to make sure the nodes we choose will collectively have the most influence. This is not an easy task, especially in realistic online social networks that can have billions of nodes.

Summary of Part IV

In this part of the book, we examined scenarios in which people are influenced by others' decisions, whether that was to view a video or purchase a product. We looked at how population-based models like the information cascade shed light on why some YouTube videos become viral while others do not. Then we brought network topology into the picture and investigated how we might identify and leverage influential people on social networking platforms like Facebook and Twitter.

The overarching theme here was the opposite of the principle in part III. In the presence of social influence, people's actions become dependent on one another, shattering the fundamental assumption behind the wisdom of crowds. When crowds follow their interdependence they can be leveraged to spread information to the masses, even if it's incorrect.

PART V
DIVIDE AND CONQUER

In our exploration of the first four networking principles, we referred to the Internet many times. Most of the topics discussed in the previous chapters rely on its existence. When it was mentioned, you probably wondered, what exactly is the Internet? How is it designed, built, and managed? Parts V and VI of this book are largely about these questions. Our two remaining principles encapsulate some of the key aspects of the Internet.

In this part, we'll see how the Internet is designed and managed in a way that scales. With the number of connected devices and sheer size of the Internet constantly growing, scalability is imperative in almost every sense of the word. As we'll see in chapter 11, it calls for an efficient way of sharing the network resources, and a division of management responsibility—both geographically and functionally—so that the subparts can be tackled more easily. Then, in chapter 12, we will see how different subnets of the Internet handle the important task of routing messages from one point to another in a scalable manner. Divide and then conquer.

11

Inventing the Internet

In addition to being an extraordinary technology and tremendous commercial success, the Internet—a network of networks—exemplifies important engineering design philosophy. We start our journey through the next three chapters not with any particular technology, but with three important ideas in design that lead the Internet to great scalability: packet switching, distributed hierarchy, and modularization.

SHARING REVISITED

Do you remember back in part I of the book when we talked about resource sharing? We looked at different techniques for multiple access, which are methods that make it possible for users to share network resources. Transmitting at different frequencies, at different times, and/or with different codes are all popular methods for dividing a communication medium.

Over the network from your phone to a YouTube server, if we dedicate resources for you and everyone else, all the transmissions will be distinguished along their paths from start to end. **Circuit-switched** networks like this one belong to the group of networks that assign fixed portions of resources to each user.

Can you think of situations where circuit-switching might be inefficient? What about the bursty nature of data applications on the Internet, like the web, email, and file transfers? In these cases, not all network resources will be in use constantly, because these transmissions happen over a lot of short time periods. When typical users don't really require a dedicated resource, why not allow them to share these resources?

We could mix and match pieces of information belonging to different Internet sessions, and let them share the paths. This is the essence of **packet switching**: divide the information to be sent into smaller chunks called

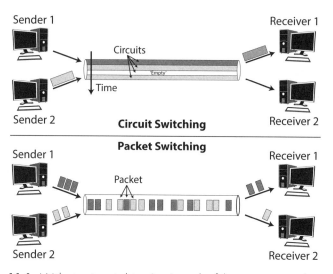

Illustration 11.1 With circuit switching (top), each of the two sessions has a dedicated resource (in this case, time) along the link. With packet switching (bottom), each session divides its messages into packets and sends whenever packets arrive. The path is shared along all timeslots and frequency bands.

packets. Each session may traverse multiple paths, and each link is shared by packets from different sessions.

Now, what exactly is an Internet **session**? It is an exchange of information, or a conversation, between two or more devices connected through the Internet. When a session is established, resources lie along the entire path from one end of the communication path (the sender) to the other end (the receiver). A session can have more than one source (i.e., sender) and more than one destination (i.e., receiver), but we focus on **unicast** sessions that have one of each.

In illustration 11.1, you can see how packet switching and circuit switching differ. Each circuit in circuit switching may occupy a particular frequency band, or a dedicated set of timeslots. In packet switching, there are no dedicated circuits.

THE INTERNET EVOLUTION

Before the 1960s, communication networking was based mostly on circuit switching. The Internet evolution, which started during the 1960s and 1970s, was marked by a shift to packet switching as the fundamental paradigm of networking. Let's take a few minutes to explore how the shift transpired. This will introduce you to the other two big ideas—distributed hierarchy and layering—behind the Internet as well.

ARPANET

In the mid-1960s, the Advanced Research Projects Agency of the US Department of Defense, or ARPA, was interested in creating a large-scale network based on packet switching. By the end of the decade, ARPA had prepared a plan that would change our lives forever.

In 1969, ARPA awarded a contract to BBN Technologies to build computers (called interface message processors) that could support their plan. Equipped with these machines, four institutions—the University of California, Los Angeles (UCLA); Stanford; the University of California, Santa Barbara (UCSB); and the University of Utah—put together the first prototype of a packet-switched network, which came to be known as the **ARPANET**. On October 29 of that year, "lo" became the literal text of the first message sent over the ARPANET, from UCLA to Stanford. The programmer was actually trying to send the word "login," but the system crashed after the first two letters due to a coding error that the team fixed an hour later.

ARPANET grew quickly. In March 1970, it reached the East Coast, at Cambridge, Massachusetts. Nine machines were interconnected by June, 13 by December, and by that next September a total of 18 sites made up the network. You can see a map of the connected hosts at that time in illustration 11.2. In 1975, the ARPANET was officially declared operational, at which time it had grown to about 60 machines.

Around this time, Robert Kahn and Vinton Cerf (see interviews with them right after parts V and VI) published a landmark paper detailing their development of a novel **protocol** for packet-switched networks. A protocol is essentially a set of rules that devices use to communicate with one another, and it specifies the common "language" they speak (we looked at WiFi random access protocols in chapter 2). This one became known as the Transmission Control Protocol/Internet Protocol, or **TCP/IP**, and gave a scalable way to connect the hosts in the ARPANET, through end-to-end control over a network of packet-switched networks. Importantly, the protocol didn't need to be modified when large numbers of different devices were added to or removed from the network, enabling interoperability and

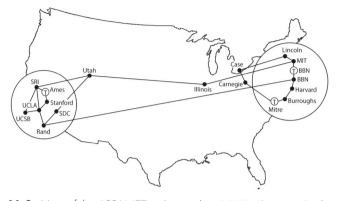

Illustration 11.2 Map of the ARPANET in September 1971, showing the first 18 hosts. MIT is Massachusetts Institute of Technology; SDC is System Development Corporation; SRI is Stanford Research Institute; UCLA is University of California, Los Angeles; UCSB is University of California, Santa Barbara.

connectivity at scale. TCP/IP would replace ARPANET's original protocol about a decade later, in 1983.

TCP/IP also gave rise to the important idea of dividing the tasks that the Internet needs to perform into different functional layers. One layer can change without affecting the operation of the protocols in another layer. As one of the fundamental ideas behind the Internet, we take a look at layering in more detail later in this chapter, in the section on modularization.

NSFNET

Up until the mid-1980s, funding and authorization issues prevented many groups from connecting to ARPANET. At this time, the US National Science Foundation (NSF) took over development, with the aim of creating an academic research network that would allow scientists to have access to large computing centers throughout the United States. From 1985 to 1995, the NSF sponsored the creation and operation of an ever-increasing network of networks, which became known as the **NSFNET**.

NSFNET was built around the three-tiered structure shown in illustration 11.3, where each type of node—campus, regional, and backbone—is itself a network. Campus networks connect together through a regional network, and regional networks are combined through the backbone,

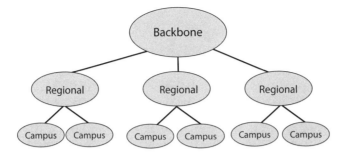

Illustration 11.3   NSFNET's three-tiered network of networks.

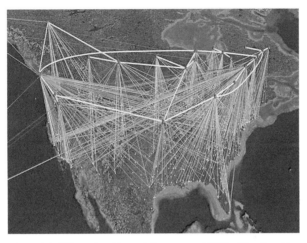

Illustration 11.4   NSFNET in the United States in 1991. The backbone nodes are drawn at the top with heavier lines, and the regional networks are indicated on the map itself.

gluing the NSFNET together. The backbone first spread across the United States and later across other countries.

NSFNET grew as more and more network providers joined. To satisfy the increasing demand, the backbone had to be improved at different times, in both size (i.e., the number of nodes comprising the backbone) and speed (i.e., of the links connecting its nodes). In 1986 it had six nodes, with link speeds of 56 kbps (or 56 thousand bits per second). By 1991, the backbone had expanded to 14 nodes, with link speeds of 1.5 Mbps (or 1.5 million bits per second). In illustration 11.4, you can see just how dense and rich the connectivity of the network had become by this time.

NSFNET was appropriated by research and education, for science and engineering. Strictly speaking, it was not allowed to be used for commercial activities (check out Q11.1 on the book's website for some of the policies). That would soon change. With demand on the rise, in the early 1990s, a number of Internet service providers, or ISPs, emerged to extend the Internet for public use (we talked about ISPs charging users for data usage in chapter 3). Commercial interests and entrepreneurial activities dramatically expanded this interconnected network of networks. By 1994, the World Wide Web and web browser had matured as a user interface, and the world quickly began developing commercial applications to run on top of the network: email, file sharing, web surfing, you name it.

In 1995, the NSFNET was officially decommissioned and superseded by the commercial Internet. Over the next two decades, the Internet would blossom into an essential part of our daily lives, with the number of people and devices connected around the world growing rapidly each year, thanks in part both to technology advances and to the scalable divide and conquer design principle. The number of people with Internet connections in their homes reached 1 billion in 2005, 2 billion in 2010, and 3 billion in 2014. The roughly 12 billion total Internet-connected devices in 2014 was an average of 1.7 devices for every person on the planet. As the **Internet of Things** (or IoT) starts to flourish, these last two numbers are expected to triple by 2020, to 33 billion devices and 4.3 per person.

With a brief history of the Internet's evolution in hand, let's now take a look at the three key architectural ideas. We start by returning to packet switching from earlier.

PACKETS VERSUS CIRCUITS

The debate between dedicated and shared resource allocation runs far and deep. There is one big advantage of circuit switching, or dedicated resource allocation in general: a guarantee of quality. Each session has a circuit devoted to it. Because of this, throughput performance (the rate of successful message delivery) and delay performance (how long it takes to deliver a message) can be guaranteed.

In contrast, sessions in a packet-switched network may share paths with one another. Any one session's traffic may also be split across different paths. Pieces of a message can arrive at the destination out of order (the receiver can reorder them), and links on the paths may get congested.

Throughput and delay performance become uncertain. In the face of such uncertainty, the Internet claims to offer **best-effort** service, meaning that it will do its best to transfer your message with high performance. Still, there's no guarantee; it's perhaps more accurately described as *no effort* to guarantee performance.

But packet switching has two big advantages. First is that it provides ease of connectivity: there is no need to search for, establish, and maintain end-to-end resources for each session. The network doesn't have to make sure that your resources are saved for you, and you don't have to wait until they are. A device can transmit at will, as long as it follows the Internet's protocols.

The second advantage is scalability. We have already seen many instances where scalability is an important property to have in networks, from supporting power control over hundreds of users in a cell (chapter 1), to running PageRank quickly on the massive graph of the web (chapter 5), to obtaining effective social learning in massive online courses (chapter 8). In this case, scalability refers to the ability of a packet-switched network to take on a large number of diverse sessions, some being short in duration, others long.

How does packet switching obtain scalability? By achieving high efficiency in the usage of network resources. What makes it highly efficient? Two "secret sauces." First is **statistical multiplexing**: many sessions can share one path and the resources along that path. Second is **resource pooling**, which is the complement of statistical multiplexing: one session can use many paths. Let's take a look at these in more detail.

Make Bigger Better: Statistical Multiplexing

In a packet-switched network, there's not just one session on each path occupying all the resources on that path. Since sessions don't have dedicated resources, they are not wasting anything when they're idle. Other sessions with demands during idle time can make use of any unused resources.

Take a look at illustration 11.5. During a few particular timeslots, Alice has much higher demand than Bob. Under circuit switching, each would be allocated different timeslots, and each would keep his/her circuit, resulting in many portions of Bob's assigned timeslots being wasted. Packet switching allows Alice to fill up this otherwise idle time. Available supply is used as long as there's unserviced demand.

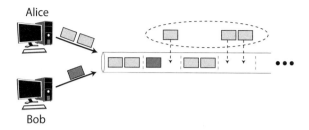

Illustration 11.5 In a circuit-switched network, portions of Bob's dedicated timeslots would be wasted during the times he has low demand. Under packet switching, multiple sessions can use the same resources, allowing Alice to fill up Bob's excess supply.

Back in part I, we used the cocktail party analogy to illustrate different multiple-access techniques. If everyone is taking turns speaking in the same room (i.e., sharing one path), what happens when one group is scheduled but has nothing to say? Under circuit switching, nothing: we would have a quiet period, during which time others wished they could be filling up the dead time. Packet switching would avoid this waste of time, through statistical multiplexing.

Make Bigger Better: Resource Pooling

One session can also use many paths to transfer information in a packet-switched network. Demonstrating the efficiency of resource pooling under bursty traffic can get pretty complicated, but the basic idea is straightforward: instead of having two sets of resources, say, two isolated links, we can combine them and use them as one large resource.

Take a look at illustration 11.6. There are two sessions and two links. At a particular time, Alice has high demand that cannot be satisfied by the available capacity on the top link alone. Bob, on the contrary, has low demand and low link utilization during this time on the bottom link. By aggregating both links into a single, larger pool, the network is able to meet some, if not all of Alice's remaining burst of demand during this time. In treating paths as shared resources rather than dedicated circuits, packet switching implements resource pooling.

Back again to the cocktail party. Suppose there are two rooms (i.e., two links), each with ten pairs of people talking. These are small rooms, so it's quite uncomfortable and congested. Then, in one of the rooms, six pairs

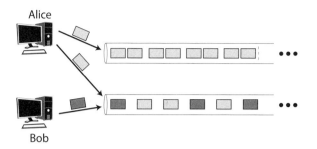

Illustration 11.6 At a particular time, Alice has more demand than the top link can handle, while Bob has much less demand than the capacity of the bottom link. Under packet switching, the two links are viewed as a single, pooled resource, allowing some of Alice's message to be transmitted over the bottom link.

Property	Circuit Switching	Packet Switching
Guarantee of Quality	✔	✘
Ease of Connectivity	✘	✔
Scalability	✘	✔

Illustration 11.7 Difference between packet switching and circuit switching in terms of three important networking properties.

decide to leave, leaving four in that room and ten in the other. Packet switching would look to alleviate congestion by equalizing demand in each room, telling three sessions in the more congested room to move to the less congested one. Circuit switching would look at these rooms as two separate resources, and would not attempt to shift sessions from one to the other.

Illustration 11.7 summarizes the three key differences that we have seen between what packet switching and circuit switching offer to a network. In the end, the abilities to easily provide connectivity and to scale up with many diverse users were more attractive for the Internet than was a guarantee of quality, although this was not completely clear until the early 2000s. Quality guarantee is certainly nice to have, but the other two properties are essential for a large and dynamic network like the Internet. Once the network has grown in an easy and scalable way, we can search for other solutions to take care of quality variation.

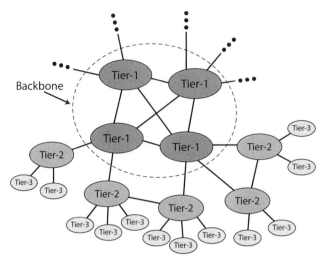

Illustration 11.8 The Internet separates Internet service providers (ISPs) into different levels. Tier-1 ISPs are all connected to one another and form the backbone of the Internet. Each of them is connected to a number of tier-2 ISPs, and these in turn connect to tier-3 ISPs.

DISTRIBUTED HIERARCHY: DIVIDING SPATIALLY

So, packet switching is great for dealing with dynamic, bursty traffic. Still, managing such a network is quite complicated, for several reasons. One obvious complication is the sheer size of the Internet; we've already discussed how big the number of people using it has become.

Related to this is the fact that the Internet has extended to almost every corner of the globe. Many different ISPs have emerged throughout the world, each owning and being responsible for a different portion of the network. Any given end-to-end Internet session may traverse links across multiple ISPs. For example, a YouTube session from Google to your iPhone will go through a wireless interface, then (if you are not connected to WiFi) probably through a few links in the core of the cellular network, followed by a sequence of even more links across different providers.

Each ISP sits at a different level of an overall hierarchy. These levels are distributed in a manner similar to the structure of NSFNET: there are three different tiers, as you can see in illustration 11.8.

In the first tier, there are a few very large ISPs called **tier-1 ISPs**. Each tier-1 has a global footprint, and can reach any other tier-1 without having to go through a lower level. This means they form a **peering** relationship and can pass traffic through one another. The full mesh of tier-1 ISPs is sometimes called the Internet backbone, like for the NSFNET. Examples include AT&T, Verizon, British Telecom, Level 3 Communications, and NTT.

At the second tier, many more ISPs with regional footprints become **tier-2 ISPs**. These may also form peering relationships among one another, but they cannot physically reach the Internet backbone without passing through their tier-1 provider(s). When a tier-2 and tier-1 ISP are connected, they form a **customer-provider** relationship, where the tier-2 has to pay a fee to the tier-1 to pass traffic through.

Finally, each tier-2 ISP provides connectivity to many **tier-3 ISPs**, forming another type of customer-provider relationship. Tier-3 ISPs take traffic only to and from customers, not other ISPs. For example, campus, corporate, and residential ISPs in rural areas belong to this group.

Each ISP and the associated administrative entities on the Internet form an **autonomous system**, or AS. As of mid-2014, there were more than 45,000 ASs throughout the Internet, up from 30,000 in late 2008. This is yet another example of how large and geographically distributed the Internet has become. In fact, dealing with Internet traffic that stays within an AS, or intra-AS, is much different from dealing with traffic that traverses two or more ASs, or inter-AS. Letting each AS manage its own traffic is a way of scaling up the Internet by distributing control spatially.

Not all the communication must pass through the entire network, either. Driven by applications in the Internet of Things and in immersive artificial intelligence, **fog** architecture, for example, is poised to distribute computations, storage, control, and communication closer to the end users.

MODULARIZATION: DIVIDING FUNCTIONALLY

The complexity of the Internet stems as much from its size as from the many tasks it has to handle. It needs to route messages through the network, control congestion, run applications, establish sessions, and perform many other functions. How can all of these tasks be managed? When engineering a complex system like this, it is natural to modularize the functions, that is, to split them into smaller pieces that can each be managed separately. Divide, and then conquer.

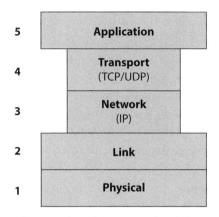

Illustration 11.9 A typical protocol stack associated with the Internet, consisting of five layers. (IP, Internet Protocol; TCP, Transmission Control Protocol; UDP, User Datagram Protocol.)

Stacking Protocols: Layering as Decomposition

Modularizing network tasks has led to what's known as a **layered protocol stack**. Each layer in a stack has a different purpose, being responsible for different sets of functions that can be managed together.

Illustration 11.9 shows a typical protocol stack associated with the Internet. There are five layers here: physical, link, network, transport, and application. They are referred to from bottom to top as layers 1 (physical) through 5 (application), each using service provided by the layer below and in turn providing service to the layer above.

Let's take a look at this stack in a bit of detail. At the bottom lie the physical and link layers, 1 and 2. The **physical layer** deals with the transmission of signals over the network medium. This medium could be copper cable, fiber cable, a wireless interface, and so on. The **link layer**, in turn, manages device access to the network medium. It acts as something like a neighborhood traffic police officer, arbitrating among parties contending for access to a street (i.e., the links). Power control and random access that we saw back in part I, for example, are functions that operate at these layers.

In the middle, we have the network and transport layers, 3 and 4:

- The **network layer** is responsible for routing hop by hop, link by link. The protocol it uses is IP. The important function of routing happens at this layer, which we discuss in chapter 12.

- The **transport layer** focuses on managing end-to-end sessions, with TCP as its dominant protocol. We look at congestion control, an important function handled by this layer, in chapter 13.

At the top resides the **application layer**, 5. This layer is the most apparent to us as end users of the Internet. Many of the networks discussed in this book are formed from different applications that we interact with every day: the web, email, mobile apps, content sharing, and so forth. Since the 1990s, one common protocol for this layer has been the Hypertext Transfer Protocol, or **HTTP**, which is the basis for the World Wide Web.

As we said, each layer provides a service to the one above and uses a service from the one below. For example, the transport layer 4 provides an end-to-end connection—running the services of session establishment, packet reordering, and congestion control—to layer 5 above, which runs our applications. In turn, the transport layer takes the service from the network layer 3 below it, including the connections established through routing.

Over the short span of the Internet's evolution, the physical medium's transmission speed has gone up more than 30,000 times, evolving from 32 kbps dial-up to 10 Gbps optical fiber and 100 Mbps WiFi. The applications running on the Internet have evolved from expert-friendly command-line tools for file transfer (which still have their time and place) to consumer-friendly sites like Netflix and Twitter. Yet the Internet itself has continued to operate through these incredible transformations, thanks in large part to the "thin waist" of TCP/IP that stays mostly the same as the applications and communication media change.

The horizontal lines in illustration 11.9 are meant to be boundaries between the layers. They are actually very complicated objects, representing limitations as to what each layer can do, what it can see, and what it is responsible for. The boundaries are not clear-cut either, because some functional overlaps occur between layers. A prime example is error control, which tries to detect and deal with mistakes in transmissions; each layer does some amount of this task. The overlaps are intentional, building in functional redundancy that helps ensure robustness while allowing evolution of the network through the layered structure.

By dividing tasks according to the protocol stack, the Internet can conquer each layer in a manner that scales as the set of functions grows.

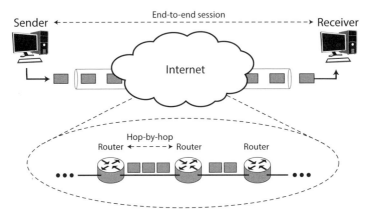

Illustration 11.10 The transport layer (layer 4; see illustration 11.9) is in charge of end-to-end session management, whereas the network layer (layer 3) is in charge of hop-by-hop management.

Connection-Oriented versus Connectionless

You may have noticed that when we introduced TCP and IP near the start of this chapter, we referred to them together as TCP/IP. Why is that, if they are used in separate layers? In fact, the initial versions of the Internet had TCP/IP as a single protocol, managing both hop-by-hop routing and end-to-end sessions. In accord with the principles of layering and modularity, in the early 1980s, TCP and IP were officially split into two parts to serve the transport and network layers, respectively.

Between these two layers, quite a few interesting architectural decisions have been made. The transport layer, in charge of end-to-end management, is **connection-oriented** in TCP, whereas the network layer, in charge of hop-by-hop management, is **connectionless** in IP. You can see this distinction in illustration 11.10. This division of responsibility implies that the network layer doesn't care about congestion or load conditions on the links. This is left to the transport layer, TCP in this case, which takes care of managing demand at the end hosts.

The difference between connection-oriented and connectionless communication can also be seen through an analogy to a phone call versus a mailed letter (illustration 11.11). If you call someone on the phone, before you begin talking with them, they have to be alerted that you are calling (the phone rings) and pick up the receiver. In this way, you are establishing

Connection-oriented   Connectionless

Illustration 11.11 *Making a phone call versus mailing a letter is analogous to connection-oriented versus connectionless communication.*

a connection with them prior to your conversation. In contrast, think about what happens when you mail someone a letter. Each intermediate post office that receives it along the way only cares about where the letter is going next, not where it came from or where it's supposed to end up. It only uses the final destination to determine the next hop on the path. During this connectionless process, the recipient may not even know the letter is on its way. We will come back to this postal system analogy in our discussion of routing in chapter 12.

The Overhead of Modularization

We use the word "message" many times when referring to something being sent over the Internet, like an email, instant message, or anything else. Technically, a **message** is the fundamental unit of data generated at the application layer. Before a message gets transmitted, each layer in the protocol stack will add its own **header** to it, so that the layers on the stack at each node in the network can interpret their corresponding headers. Starting from the top at layer 5, and making our way down to layer 1, here's how data may be **encapsulated** for transmission:

Layer 5: When an Internet user wants to send something, her device tells the application layer to generate a sequence of messages for transmission.

Layer 4: Each message is divided into **segments** at the transport layer. Each of these segments contains two things: the **payload**, which is some of the actual content from the message, and a layer 4 header, added in front of the payload.

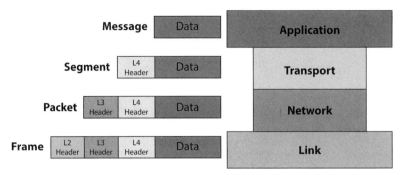

Illustration 11.12 Layers in the protocol stack add their own headers to a message, encapsulating it prior to transmission over the Internet.

Layer 3: Each segment is then passed to the network layer, which divides and encapsulates it as datagrams, or **packets**. Each of these packets has a layer 3 header in the front.

Layer 2: Each packet is further passed to the link layer, which adds a layer 2 header to form a **frame**.

Layer 1: Finally, the frames are all passed to the physical layer for transmission as bits.

You can see this process in illustration 11.12.

Different devices in the network—computers, routers, modems, servers, and so on—run different subsets of the layered protocol stack. Each will **decapsulate** (i.e., decode) and read the header information associated with the subset that it runs. If it's an intermediate node along the path, it will then encapsulate those layers again and send the information on. A few important cases can be found in illustration 11.13:

- Computers and servers, being end hosts, process all five layers.
- **Routers**, as network-level devices, go up to the third layer. This requires them to have, as well as to process, IP addresses.
- **Switches**, as link-layer devices, only process up to layer 2. They do not have or process IP addresses.

Encapsulation may seem unnecessarily redundant to you. You are right in that the process creates control **overhead**, one of the many kinds of redundancy and overhead in a layered architecture, because there's a significant amount of data being sent over the network that is not actual content. So, why do the layers have to add headers in the first place? It's a way

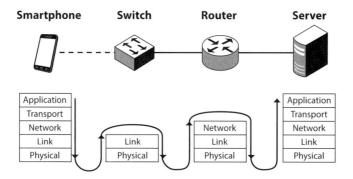

Illustration 11.13 Different network elements process different layers in the protocol stack.

of distinguishing among, and providing information about, different transmissions in a packet-switched network, like where they came from, where they are going, and how long they have been in transit. As an example, the layer 3 header contains the source and destination IP addresses, which (as we shall see in chapter 12) is essential for the Internet's task of routing. Again, by allowing everyone to share the same network resources, packet switching can obtain much higher efficiency than circuit switching, with the caveat that we need to distinguish among sessions within the transmissions themselves.

Packet switching, distributed hierarchy, and layering are three fundamental concepts behind the Internet. They have allowed the Internet to scale effectively with higher demands, broader geography, and more functions to handle. But we have only just begun our exploration. Many tasks are involved in managing an ever-expanding network of networks: we have to figure out how to get from point A to point B, manage congestion along links, and so on. Our discussion on autonomous systems from illustration 11.8 is a good segue to the next chapter, which will focus on intra-AS routing: routing within an AS.

12

Routing Traffic

How does traffic get from one place to another through the Internet? A short answer is that the network contains devices called routers that direct packets of data where they need to go. A word we use in everyday speech, **routing** on the Internet is similar in its objective to determining which route you will follow when driving somewhere. Before we dive into a method for determining packet routes, let's take a look at the main ideas.

THE INTERNET'S "POSTAL MAIL SERVICE"

As we've seen before, transportation networks can offer useful analogies to communication networks. The postal mail service gives an interesting one for Internet routing. To route from sender to receiver, we need three main features: addressing, routing, and forwarding. Sometimes these three terms are lumped together in an informal conversation, but they are indeed distinct steps.

Addressing

When you send a letter through the mail, you have to address the envelope (illustration 12.1). Otherwise, the postal service would have no idea what to do with the letter. You put the recipient's street address, town, state, country, and zip code on the front, which tells the post office where it should have the letter taken to. You put your address there too, so that it's clear where the letter came from, both for the recipient's benefit and in case there is some problem along the way.

Your recipient's postal address gives her a unique label that nobody else in the world shares (except people living in the same house), so that there's no ambiguity as to where you want the letter to go. This is how addressing works for the Internet, too: we attach a unique label to each node in the

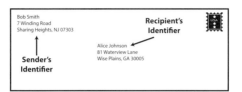

Illustration 12.1 Addressing an envelope to be sent through the postal system is analogous to addressing a message to be sent through the Internet.

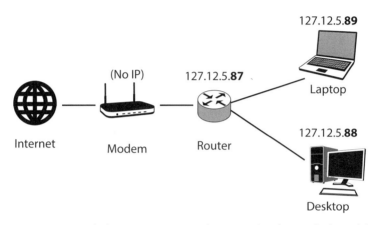

Illustration 12.2 Network devices are assigned unique identifiers called IP addresses.

network, so that we can identify the sources and destinations of messages. In particular, each network device is assigned an Internet Protocol address, known as an **IP address**. IP addresses are typically represented as decimal numbers separated by dots, for example, 127.12.5.88. The decimal numbers can be between 0 and 255.

In illustration 12.2, you can see a few common Internet devices in a home, and which of those are often assigned IP addresses. Remember from chapter 11 that some devices, like modems, don't use the Internet Protocol to communicate and therefore don't need to be assigned IP addresses.

There are two versions of IP, version 4 and version 6. **IPv4** uses four numbers for its addresses (e.g., 127.12.5.88), allowing for more than 4 billion possible addresses that can be assigned. As of early 2011, the Internet had reached the point where 4 billion was no longer enough. The addresses in **IPv6** were designed to use the equivalent of 16 numbers for addresses, giving about 4 billion $\times$ 4 billion $\times$ 4 billion $\times$ 4 billion possibilities. This might sound unnecessarily large (as the 64-bit YouTube view counter in chapter 9 may have), but with the proliferation of Internet-connected devices, it has proven to be a wise choice, especially considering how these addresses will be allocated in the Internet of Things.

Let's dig a little deeper into the postal system analogy. Similar to how the zip code of the mailing address gets our letter to the specific town or city, the **prefix** of an IP address gets our Internet message to the destination router. For an IPv4 address, the leftmost three decimal numbers might indicate the prefix (though it could be longer or shorter). It is specified with a forward slash: 127.12.5.0/24 means the prefix for this address is 127.12.5. Why would 24 refer to the first three numbers? Usually, lengths in IP addresses are given as the number of bits used to represent them. Each decimal number in this case takes up 8 bits, so three decimal numbers take up 24.

Using the destination IP, routing will bring the message to the group of devices that has the same prefix. This group is known as the **subnet**. The numbers after the prefix give the specific **host identifier** within the subnet. For example, the devices in illustration 12.2 are each in subnet 127.12.5.0/24 and have host IDs of 87, 88, and 89. In the mail system, we could say that all houses sharing your zip code are in your "subnet," while your street address serves as your "host identifier." For more on subnets, prefixes, and host addresses, check out Q12.1 and Q12.2 on the book's website.

An end-user device typically won't have a fixed, static IP address. Usually, one is automatically assigned and leased for a given amount of time. This service is offered by a Dynamic Host Configuration Protocol, or **DHCP** server, providing pertinent IP addressing information to the device (illustration 12.3). The DHCP server keeps track of what IP addresses are free to take. Your device contacts it and obtains a currently unused IP address for a finite lease time. When the lease expires, it can be renewed; otherwise the server will return it to the address pool, so other devices can use it.

Sometimes, the IP address given to your device in your local network is different from the one seen by the rest of the Internet. The address of your laptop within your college campus, for example, may be different from how it appears to the outside world. A network address translation, or **NAT** router is in charge of translating back and forth, so that people from outside your local area can address your device based on your public IP (illustration 12.4). You can think of NAT like a mailroom in a company building: when a package is sent to an individual in the company, the postage system

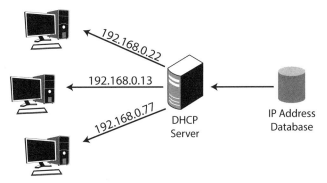

Illustration 12.3 A Dynamic Host Configuration Protocol (DHCP) server is responsible for leasing Internet protocol (IP) addresses to devices. It keeps track of the available addresses that haven't been assigned yet in a local database.

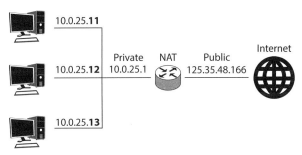

Illustration 12.4 Sometimes, your public Internet protocol (IP) address is different from your private IP address. A network address translation (NAT) router is in charge of converting between the two.

uses the building address (i.e., the public IP) to get it there. Then, a sorter (i.e., the NAT router) in the mailroom uses the person's name (i.e., the private IP) to determine whose mailbox to place it in or which desk to send it to.

Routing

With addressing in order, the next step is to decide on the path that a message will take to its destination.

In the postal system, routing decides ahead of time which intermediate cities the mail will go through to reach, say, Miami, Florida, from Princeton, New Jersey. From the local post office in Princeton, the mail might be transferred to a large, regional office somewhere in New Jersey, then down to a large office somewhere in Florida, then to the recipient's local post office, and finally to the recipient's door.

In the Internet, the paths that messages take vary, depending on the routing method used. There are two classes of methods:

- Inside an autonomous system (AS), **metric-based** routing is used. With metric-based routing, the objective is typically to find the shortest or least-congested path to the destination.
- In between ASs, routing is instead **policy-based**. For example, one AS might suspect that another has a lot of hackers, and will want to avoid routing packets along any path that goes through that AS.

Inter-AS routing is very different from intra-AS routing. Border Gateway Protocol, or **BGP**, is the dominant protocol for routing between different ASs. It glues the Internet together. For within an AS, there are two main flavors of routing protocols:

- Routing information protocol, or **RIP**, where each router collects information about the distances between itself and other routers; and
- Open shortest path first, or **OSPF**, where each router tries to construct a global view of what the entire network topology looks like.

In this chapter, we go through the main ideas behind RIP, skipping some intricate details of the most popular type of routing: OSPF.

Forwarding

Back to our example of sending a letter from New Jersey to Florida. When the post office employees in New Jersey get the letter, what will they do? They will look at the zip code on the envelope, and see it is headed to Florida. They probably don't care about the specifics, like what town or city, just that the letter has to get to some regional office in that state. So they will put the letter on a plane destined for Florida. Then, once the letter arrives there, someone will look at the zip code again and know they have to get it to Miami. So they will put it in the mail truck that is headed over there. Finally, once the letter is within the zip code in Miami, the local post office will look at the home address and deliver it to the recipient.

In the Internet, the action of **forwarding** occurs each time a packet of data is received at a router. When one arrives, the router looks at the destination IP address written in the packet, figures out where it has to go, and sends it off to what becomes the next hop on the path. Then the next router receives the packet, looks up the destination, forwards the packet along, and the process continues in this hop-by-hop fashion. It's a connectionless process, as we talked about in our discussion on modularization in chapter 11.

Depending on how far away the destination is, the first routers on the path will probably only care about the prefix of the address. In other words, they first forward based on the destination subnet, similar to how the postal system first forwards toward the destination zip code. Once the packet reaches the destination subnet, the final routers will forward based on the host identifier to get to the specific device.

Physically, how does forwarding occur? A router is connected to other routers in the network by links. When a packet arrives at an incoming link, the router will move it to an outgoing link. Inside the router is hardware built to perform this function as quickly as possible, connecting the input and output ports that manage access to the links.

How does a router determine which output link is the right one for a given message? This information is kept in a **forwarding table**, which maps destination IP addresses to output links, as you can see in illustration 12.5. Using its forwarding table, the router can look at the destination IP address, find its entry in the table, and choose the corresponding outgoing link, like using a phone book to look up someone's number from their name. Each entry in the table may represent a large range of addresses, for example, 10.1.2.1 and 10.1.2.10 could both go to Link 3 in illustration 12.5.

The forwarding table needs to be built, maintained, and updated. The way that this is done depends on the routing method used by the network. Many different schemes have been proposed for doing this over the years. We turn next to understanding how these tables can be built by looking for the shortest paths through the network.

FINDING THE SHORTEST PATHS

So, what is the overall objective behind routing? To get from one point (the source) in the Internet to another (the destination) in the best way possible. The routers are the intermediate nodes that pass the messages through the network.

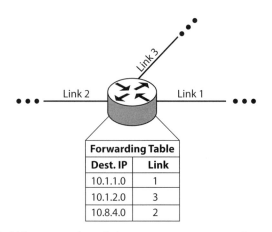

Forwarding Table	
Dest. IP	**Link**
10.1.1.0	1
10.1.2.0	3
10.8.4.0	2

Illustration 12.5 When a packet of data arrives at a router, the router will look up the destination IP address in its forwarding table to determine which outgoing link to forward the packet to.

We have seen many different graphs in this book, from webgraphs in chapter 5 to social graphs in chapters 8 and 10. For routing, we now introduce another one: the graph of routers.

The Graph of Routers

Take a look at illustration 12.6. In a routing graph, the source's job is to determine which of its neighbors (here A, B, or C) to forward the message to. That chosen node will in turn forward to other, intermediate routers, and so on, until the message arrives at its destination. Again, forwarding is done one **hop** at a time, where one hop is one link.

There can be multiple ways to get from the source to the destination. How do we determine which path is the best? Usually, we want to take the one that has the lowest cost. Each link in a routing graph represents a physical connection from one router to another, and each has a different cost associated with it. Often, this cost is related to the distance between the two routers that the link connects; for example, the one between two routers in the same room might be less than the one between two in different buildings.

What type of graph should we use to include the costs? A weighted graph: similar to how we weighted hyperlinks by importance scores in chapter 5, we weight router links by costs. In illustration 12.6, the numbers on the

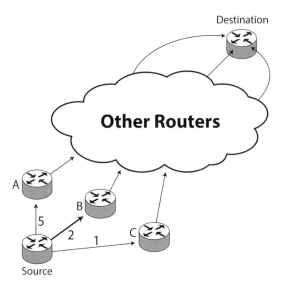

Illustration 12.6 Routing concerns the transfer of traffic from a source to a destination. To do this, each node needs to know which of its neighbors is best to forward the traffic to for a given destination.

links indicate that sending from the source to A has a cost of 5, whereas sending to B costs 2. Also, the links in a routing graph are directed: if A can forward to B, that doesn't mean that B can forward to A. If they can forward to each other, the cost need not be the same in each direction, either.

The task of finding a minimum-cost path from one node to another is a well-known problem in graph theory called the **shortest-path problem**, as people often think of a link's cost as its distance. "Shortest path" becomes "fewest hops" when the link weights are all equal.

You can see a simple example of the shortest-path problem in illustration 12.7, with four routers and four links. If A wants to send to D, it can either forward to B or to C. The cost along path (A, B, D) is $2 + 4 = 6$, and that along path (A, C, D) is $3 + 5 = 8$. Since (A, B, D) is the shortest path, A should forward to B (and then B to D).

The Bellman-Ford Algorithm

How can shortest paths between routers in an autonomous system be discovered? It needs to be done in a way that is scalable with the number of

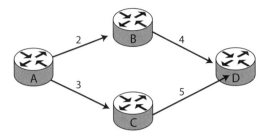

Illustration 12.7 Path (A, B, D) costs 6, and path (A, C, D) costs 8. If A wants to send to D, it should forward to B, which will in turn forward to D.

nodes and links in the network and that can adjust to any changes in the topology relatively quickly.

The shortest-path problem has been studied extensively since the early 1950s. Several famous algorithms have been developed to solve it, such as Bellman-Ford, Dijkstra's, and A* Search, each with their own advantages and disadvantages. In this chapter, we focus on the **Bellman-Ford algorithm**, because it's simple yet elegant and illustrates the fundamental principles behind routing algorithms. It also leads to implementations used in some famous routing protocols, such as the original ARPANET.

Bellman-Ford gets its name from the American mathematicians Richard Bellman and Lester Ford, Jr., who published the algorithm in 1958 and 1956, respectively. Bellman is well known more generally for having introduced in the 1950s the method known as **dynamic programming**, important for solving complex problems in mathematics, computer science, economics, and other disciplines by breaking them down into smaller, simpler subproblems that can be solved more readily (fitting the overarching principle of "divide and conquer"). The equation in dynamic programming that relates a given problem to its subproblems is usually called the Bellman equation.

Like distributed power control in chapter 1, the Bellman-Ford algorithm is an iterative procedure that repeats over and over, until we can be sure that it has finished. In each iteration, it finds a path from source to destination that may be a shortest path and then uses this information in the next iteration to see whether it can find an even shorter path. The first step finds a shortest path by using only one hop, the second step for up to two hops, the third step for up to three hops, and so on. Using more hops

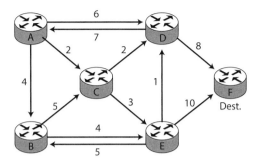

Illustration 12.8 Example graph of six routers, with link costs indicated.

adds more possibilities, so the cost will decrease (or stay the same) each iteration along the way.

Let's use the router graph in illustration 12.8 as an example for walking through the Bellman-Ford algorithm. There are six routers, A–F, and the link costs are indicated. The objective is for routers A–E to figure out their minimum-cost paths to get to the destination F.

First Step

For the first step, we just need to know which nodes have one-hop paths (i.e., a link) to F and which do not. A, B, and C cannot get there in one hop, because none of them has a direct link to F. But D and E can: their direct links cost 8 and 10, respectively. So, the shortest paths and associated costs for one hop are:

$$D : \quad \text{Path} = (D, F), \quad \text{Cost} = 8$$
$$E : \quad \text{Path} = (E, F), \quad \text{Cost} = 10$$

Illustration 12.9 summarizes first step, where the links are highlighted and the costs to F indicated.

Second Step

Moving to the second step, the question to ask is: what are the shortest paths from each node to F using at most two hops? The Bellman-Ford algorithm uses the information from the first step to answer this. After all, if a node has a neighbor it can forward to that could reach F in one hop, then this

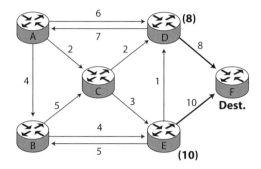

Illustration 12.9 Shortest paths and costs for one hop. D and E can each reach F in one hop, while A, B, and C cannot.

node must be able to reach F in two hops through that neighbor. To determine the total cost of this path, we can add the cost that it takes for the node to get to the neighbor to the cost that it takes for the neighbor to get to the destination.

Let's start with A. It has three outgoing neighbors: B, C, and D. Neither B nor C can reach F in one hop, so they aren't helpful right now. In contrast, D can reach F in one hop with a cost of 8, and it costs 6 for A to forward to D in the first place. Therefore, A can reach F in two hops through path (A, D, F), with cost $6 + 8 = 14$. This is A's shortest (and only) two-hop path to F.

How about B? It has two outgoing neighbors: C and E. C can't reach F in the first step, but E can get there with a cost of 10. Since B forwarding to E costs 4, B can reach F in two hops over path (B, E, F) with cost $4 + 10 = 14$.

Moving now to C, it has two outgoing neighbors: D and E. They are both able to get to F in the first step:

- Since D's cost to the destination is 8, the total cost through D is $2+8 = 10$.
- Since E's cost to F is 10, the total is $3 + 10 = 13$.

Since 10 is "shorter" than 13, C will choose (C, D, F) as its two-hop path to get to F.

What about D? This node has two outgoing neighbors: A and F. A cannot get to F in one hop, so forwarding directly to F is still its only choice. Actually, in a situation like this where the router can forward directly to the destination, would it ever want to send somewhere else? Possibly. Remember,

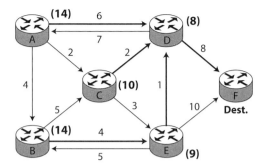

Illustration 12.10 Shortest paths and costs for two hops. All nodes can reach F now. B still thinks the total cost through E is 14, because it doesn't yet have the updated cost from the second step.

the shortest-path problem is considering minimum cost, not minimum hops. It is possible for there to be intermediate links with lower total cost.

To this point, let's consider E. D and F are its two outgoing neighbors:

- Forwarding directly to F costs 10.
- D reaches F with a cost of 8, for a slightly lower total of $1 + 8 = 9$.

So, E will choose (E, D, F).

To summarize, the discovered paths after the second step are:

$$A: \quad \text{Path} = (A, D, F), \quad \text{Cost} = 14$$
$$B: \quad \text{Path} = (B, E, F), \quad \text{Cost} = 14$$
$$C: \quad \text{Path} = (C, D, F), \quad \text{Cost} = 10$$
$$D: \quad \text{Path} = (D, F), \quad \quad \text{Cost} = 8$$
$$E: \quad \text{Path} = (E, D, F), \quad \text{Cost} = 9$$

as you can see in illustration 12.10. Compared with illustration 12.9, A, B, and C can now reach the destination, and the path chosen by E has changed.

Let's take a moment to analyze what Bellman-Ford is doing to discover shortest paths. At each step, a router looks at each outgoing neighbor and says to itself, "it costs me w to forward to you, and it costs you x to get to the destination, so it will cost me $w + x$ to get to the destination in one more hop if I forward through you." The router will choose the lowest total cost among all its neighbors. You can see this idea in illustration 12.11.

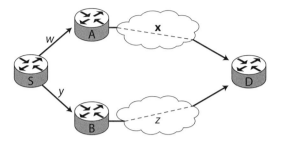

Illustration 12.11 At each step of the Bellman-Ford algorithm, the source checks among its outgoing neighbors to see which one offers the lowest total cost to the destination. To do this, it must know two things: the cost to get to its neighbors (here, w and y) and the total cost from each neighbor to the destination (here, x and z).

The source (S) has two outgoing neighbors, A and B. A can reach the destination (D) in, say, six hops with a cost x, and the cost of link (S, A) is w. So, S can reach the destination through A in seven hops with a cost $w + x$. Similarly for B, the cost will be $y + z$. Whichever of these costs is smaller is what S will choose at this step.

Third Step

Continuing now for three hops, what is the situation for A? Its three neighbors can reach the destination in at most two hops:

- B's cost to F is 14, so the total cost through B is $4 + 14 = 18$.
- C's cost is 10, so the total is $2 + 10 = 12$.
- D's cost to the destination is 8, so the total is $6 + 8 = 14$.

Since forwarding to C gives the lowest cost of 12, A will choose (A, C, D, F).

How about for B? B can forward either to C, with a total cost $5 + 10 = 15$, or to E, costing $4 + 9 = 13$. B will stick with E, with B's path now reflecting E's change from the second step.

You can check that C, D, and E's paths will not change from last time. The discovered paths at this point are:

$$A: \quad \text{Path} = (A, C, D, F), \quad \text{Cost} = 12$$
$$B: \quad \text{Path} = (B, E, D, F), \quad \text{Cost} = 13$$
$$C: \quad \text{Path} = (C, D, F), \quad \quad \text{Cost} = 10$$

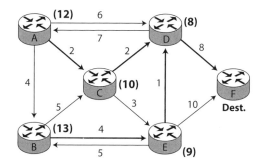

Illustration 12.12 Shortest paths and costs for three hops.

$$D: \quad \text{Path} = (D, F), \quad\quad \text{Cost} = 8$$
$$E: \quad \text{Path} = (E, D, F), \quad \text{Cost} = 9$$

The result is highlighted in illustration 12.12.

Fourth, Fifth, … Steps

There's actually no need to continue to four, five, or more hops, because the algorithm has already determined the shortest paths in this example. In general, the Bellman-Ford algorithm may need to go for up to the number of nodes in the graph before you can be sure it has finished. For more information on this, check out Q12.3 on the book's website.

Even so, in a real scenario of routing in an autonomous system, we usually do not know how many nodes there are in the first place. Instead, iterative shortest-path algorithms have to rely on the lack of changes in the shortest paths discovered from step to step to determine when it's safe to stop running. Stopping can only be temporary, though: the network configuration may change over time, impacting the link costs, the number of nodes, or both, in turn affecting what the shortest paths are at any given point. The algorithms need to be run regularly to keep their path-cost calculations up to date.

PASSING MESSAGES TO DISCOVER SHORTEST PATHS

Routing on the Internet is implemented by forwarding, which occurs one hop at a time. Each router's forwarding table will only contain information about where a message should be sent next, as in illustration 12.5. In our example (illustration 12.12), when a message arrives at router A destined for F, A only needs to know that it should forward to C. It doesn't have to concern itself with what path the message will take after that. Similarly, B only knows to forward to E, C to D, D to F, and E to D. This is a pretty amazing property of distributed routing!

Building forwarding tables in practice requires one additional, important step. The tables need to account for the fact that each router only has a local view of the network, meaning that it only knows the existence of, and path-cost information for, its neighbors. The shortest paths need to be discovered in a distributed manner, with each router building its forwarding table from local information.

How can this be accomplished? Through message passing. Routers have to send messages to their neighbors that indicate the path-cost information they have discovered and continually update their own tables based on the messages they are getting. Each message is a short summary of all the destinations the router can reach and the total cost of the path to each of those destinations.

Message passing is done locally, neighbor-to-neighbor. Through this process, routers are getting all the information they need about end-to-end shortest paths without actually knowing what those paths are. After a few steps of message passing for our example network in illustration 12.12, A will know that its shortest path to F is through C. It won't know that C will in turn forward to D (as C won't know that D will forward to F), and for purposes of forwarding in the Internet, it won't need to know either.

An application of Bellman-Ford combined with message passing is how the intra-AS routing protocol RIP mentioned earlier is implemented. One of the oldest methods, it is still in use today, with the one-hop message passing between routers making it reasonably easy to implement. The other intra-AS protocol that we mentioned, OSPF, has gained much popularity over the past decades. While RIP only keeps information about neighbors, under OSPF, each router actually tries to construct a local view of the network, including the state (e.g., cost) of each link. As a result, OSPF is a type of **link-state** routing, whereas RIP is **distance-vector** routing. OSPF is preferred especially in larger networks, where link conditions may change quickly.

Routers passing messages to build forwarding tables is another example of distributed algorithms and coordination in networking, one of the major themes in this book. It allows for scalable management as a network expands and changes over time. Power control and carrier sensing from part I are other distributed procedures we've already seen, and we'll see yet another when we discuss congestion control in chapter 13. Remember: the operation of these algorithms can be contrasted with the centralized nature of the ranking and recommendation procedures that we discussed in parts II and III.

Summary of Part V

The Internet is a constantly expanding network of networks, remaining scalable both functionally and geographically through the principle of divide and conquer. We just saw three of the fundamental concepts behind its design: packet switching, in which resources are shared instead of dedicated, distributed hierarchy, in which control is spread across different segments of the network geographically, and modularization, in which tasks are divided into different functional layers and managed separately. We also went into detail about routing, the important task of getting traffic from one place to another in the Internet. This, too, occurs in a scalable manner, with shortest paths discovered through distributed message passing between routers, and is yet another example of distributed algorithms and coordination in networks.

A Conversation with Robert Kahn

Robert Kahn is recognized as one of the ``fathers of the Internet.'' He co-invented TCP/IP with Vint Cerf.

Q: Bob, do you think the Internet could be where it is today without TCP/IP?

Bob: I consider TCP/IP to be the lingua franca. It is the set of protocols and procedures that cause the different components that comprise the Internet to be able to link together. So could one have a different set of protocols and procedures that make it possible? Yes. My guess is that, for the same kind of architectural robustness, you probably would need something fairly close to what we did before in TCP/IP, but it could be somewhat slightly different. You need something like that to allow the pieces to play together, because otherwise you've got an end-by-end interconnectivity problem where all the interfaces could potentially be quite different. So I think you probably wouldn't have an Internet today without either the protocol that Vint Cerf and I did, or with something very similar.

Q: But looking back, it wasn't always obvious that's where we were heading. What do you think made TCP/IP eventually the glue that has held every part of the network together in an interoperable way over the course of the last 40 years?

Bob: Well, the world has really evolved quite a bit in the last 40 years, even the basic notions of what might comprise the Internet have changed somewhat, but still, you have a basic need for interconnectivity. And at the time that we did that work, which was back in the early 70s, there really weren't many alternatives on the table. The Europeans eventually decided to do something similar, and they converged on an IP approach that was almost the same as ours, and eventually merged, so that the two parties adopted the same basic IP strategy, and there were a variety of end-to-end protocols. The Europeans called theirs, I think, TP0, TP2, and TP4, and eventually those didn't take hold, simply because too many had already adopted TCP before that.

So I think it was more a matter of building critical mass, that the growth of the Internet was incremental. What people wanted to do was to connect a body of participants that they could interact with, and most of them—by the time there were even fragmentarily different choices available—were all basically committed to TCP/IP. And even when the international standards came out, nobody was willing to make the time and effort to switch over, because there was no cost benefit for doing so.

Q: TCP/IP was really one protocol at the beginning in 1974, and then it later split into two layers. How would you describe that evolution from one protocol into two layers?

Bob: Well, the problem that people perceived really was driven by the need for real-time communication, and the first out of the box was the program that I started at DARPA on packet voice. So the idea [was] that you could take an analog voice stream, digitize it, chop it into little pieces—we called them parcels—and send them independently either in packets or quantified in some way into larger aggregations through a network, and be able to reconstruct a continuous voice stream at the other end from the digitized voice. That was the question. We were able to show that it was possible to do that.

Now, with the TCP in the original ARPANET protocols, every time a message went through, you'd have an acknowledgment go back before you could send the next one, and so delays were possible. The net effect was, you could have breakup in the voice. Well, we disabled that with the help of BBN [Technologies] by introducing a new type of packet that didn't require acknowledgment. We called those type-3 packets in the ARPANET.

But when we got to the Internet side of things, the end-to-end protocol was the TCP protocol, which had IP bundled inside of it, so it's composite. The idea that I had was that an application program would be talking to the TCP program and explaining what it needed. So if it was, let's say, a voice application, the application would say: "Send me the packets as they come in, but don't bother sending me any packets if they're more than, let's say, 20 or 40 or 60 milliseconds late. I'm not going to play them out, because it's too late for me to use them." And my thought was that the TCP/IP program could decide what to pass on to the program, depending on what its needs were. Well, it turned out that it was really too hard a problem to get every application that was already built to be able to be recoded, to be able to explain what its needs were, to then have the TCP program be able to accept inputs from all of those programs. So a simpler choice was made to just say, let's split it into a part that's an end-to-end part only and a part that's just an unreliable delivery piece, so it'll deliver the packets as they show up and let the end-to-end part figure out how to put them back together again in some form.

So that's what led to the split. It was sort of the perceived need to do that, and I guess a lot of people didn't recognize that there were other alternatives. But looking at it from hindsight, what I thought I had proposed was a much better long-term strategy, but it was not really a very workable or implementable short-term strategy. That's why we went with IP, and I had to kind of agree with all in the final analysis.

Q: That's a very interesting history there. For the routing part, I'm wondering if you have some favorite analogy that you use to explain to others about how routing on the Internet works.

Bob: What I would be more inclined to do is explain how routing tables might work. Namely, something might come in from one line into a node. That node could be a packet switch, or it could be a gateway between nets, and somehow you look up and pick out of the incoming packet where it's going, see what the table says about which of the output lines to send it out, then you put it out that line. The tables basically get incremented by updates from the neighbors. Basically that particular switch or gateway decides who is closest to the final destination and usually tries to send it to that place. You might have tables that are more comprehensive, global tables that represent the whole net (if you knew the whole topology), and then you could make an optimized decision about how to route it. Or you might have information about traffic flows on the different lines [when] trying to send it down paths of least congestion, rather than just closest connectivity.

That's what I would do if I were going to try and describe how routing works. But that is very dynamic and adaptive, so it doesn't have to be that you pick all the routes from places up front and just stick to that [strategy], although you could.

Q: What would you have done differently or in addition if you could go back to, say, 1974?

Bob: Well, most of the steps that we were taking were pretty much self-evident incremental steps, except for the overall Internet architecture, which we had to invent, essentially. How would all of these things work together? So we had to create the notion of IP addresses, we had to create the notion of gateways or routers, protocols that communicate the traffic and so forth, and a lot of intermediate things like gateway-gateway protocols, how to advertise routes or border gateway protocol, and so forth.

I can tell you the things that I wish had gone better. I wish that we had made more progress on security in the early days. But remember, we tried to do that. It wasn't like we ignored it, which a lot of people think we did. But rather, in order to really get security in the Internet in a way that would have been meaningful, we had to get certain people with expertise in that area to work with us. They weren't even sure that this whole technology was going to be useful or lead anywhere. It certainly wasn't commercial at the time at all. And they had so many things on their plate that they really just didn't have the time or energy to pay attention. So we did some small incremental steps. We created original private line-interfaces, which you can look up on the Internet and find out about, but they were basically red and black processors connected with an encryption device in the middle. We actually tested those, and we had ways of circumventing the encryption for purposes of selective addressing. And I think we did it with one site initially, then we did it with 32 sites. So there were incremental improvements, but I wish we had done a better job back then.

The second thing that we really didn't appreciate when we started, and the architecture allowed for it, is that we were assuming we'd have lots of different capabilities out there, and we assumed we'd have a small number of networks. Remember, the networks back then were all large, generally like ARPANETs or an AT&T network or a defense net, or something that was kind of wide area. We were assuming there would be a small number of them—maybe 4, 8, 16, some number like that around the world—and very quickly we had thousands as the Ethernets hit the stage. The net effect of all of that was that the initial addressing that we had, which assumed that we'd have 32 bit address space and we would allot 8 bits to the network and 24 bits to the eventual machine on that network, was pretty soon overtaken by events. And so we ended up in real-time having to kind of reinvent how to deal with that. I think probably what we should have done early on, had we understood how much impact it would have, is start out with something like 128- or 256-bit addresses, and we wouldn't have had to go through all the trauma of doing what's now called IPv4 to v6 transition.

I've been working for years on how to reinvent the Internet around the notion of managing information, rather than just moving bits, and what

I've been working on is this thing called the Digital Object Architecture, which is getting quite a bit of traction around the world. Not so much in the US, because there's so much focus on the web, but around the world there's a significant amount of interest. I would have done that 40 years ago if I had only thought about it.

Q: Would you describe a bit about the Digital Object Architecture?

Bob: Well, what we do in that architecture is start from the notion that everything we're dealing with is a digital object, and a digital object is either a sequence of bits or a set of sequences of bits, and it's got a unique persistent identifier. And if you're talking about information, that information could be in the form of the digital object, but you could also have a digital object that represents an individual. In fact, if you resolve the object, you wouldn't get the individual but you'd get information about the individual, like maybe their public keys or where they can be contacted that day, or anything else they wanted you to know about them—maybe even their email address or whatever. And so we assume that, within the network environment, every single resource that you care about has its own identity, and therefore if you are trying to interface with something, you can find out who you're interfacing with by just having a challenge/response interaction using public keys. So digital objects have unique persistent identifiers—we happen to call them handles, but they really are clear identifiers of what you're talking about.

For example, in today's world, if you use, let's say, a URL to access a file on a given machine and you want to preserve that for history—maybe it's government information or maybe it's corporate information—in 100 years you want to be able to come back and get that information, most likely that URL won't work anymore. The machine will have gone away, the name of the company may have changed, or the information may not be in that file name. But if you gave that information a unique identifier such that no matter where that file was, as long as somebody managed it, you could resolve that identifier into information—I call it state information.

In today's scientific literature—and I'm sure you read IEEE transactions or ACM journals or any of the traditional scientific journals—they all use this system. This architecture is all used there. We call it the handle system, which has the ability to store that state information about the objects being identified, so you can then get it back. It might tell you where to go to access it if it's available. You can move the object around. It can be in printed references, it can be in electronic storage systems, but you have to go back and change the identifiers even if you move the information or change the underlying technology base. The identifier will get you there if somebody's managed it. And, of course, you have to be allowed to access it. There might be charges somebody wants to invoke, there might be firewalls against getting through to certain information, but in principle, the architecture allows it.

So there are three components to the architecture. One is this resolution system I talked about, which we call the handle system. It takes an identifier in and gives you back essentially the state information about the thing being identified. [Second,] repository technology allows you to deposit digital objects and then access them based solely on their identifiers, so they could be stored behind the scenes anywhere—in thumb drives, RAID arrays, or just cloud services—any method you want to use to store them. And it's all invisible to the user, because it's behind the walls of that repository. And then finally, registries—we call them DO repositories and DO registries—basically store metadata about the objects, [allowing] you to browse or search the registry and [returning] identifiers to you when you're all done. Recently, we've taken two of those components, the registry and repository, and merged them into a single turnkey system that will do repository functions or registry functions, because it turns out the repository needed a registry just to know what was in the repository and the registry needed a repository just to keep its metadata records. We've now produced that, and you can find a version on the Internet under the URL cordra.org.

So that's it in a nutshell. And the interesting thing about that architecture is, anybody who adopts it, independent of the technology they use behind the scenes—just like if you use IP addresses, you can have any computers behind it and whatever—you get interoperable systems, and you get around all the issues associated with the size of things. You can have repositories in the loop here that are essentially accessing these digital objects and storing them for later presentation. It comes with built-in security, because the resolution system can store public keys in it. It therefore allows you to invoke all the functions of the PKI infrastructure without any extra cost. And I think it's really a very good model for managing information going forward, because it not only keeps the information together but it's [also] its own intrinsic cataloging system.

Q: How would an Internet that operates on digital objects look different from one that operates on bytes?

Bob: I think it wouldn't affect the underlying pieces of today's Internet necessarily, because you can take a repository and connect it to today's Internet. You can take the registry or the combined version, the handle system. They just simply make use of everything in today's Internet, and if you had something that could do that, it probably could make use of that as well. So it's not one or the other, they make use of both. If it's a digital object and it's going over an Ethernet, it's going to be treated one way. If it's over a token ring, it might be treated slightly differently. The underlying communications technology will do what it does. It may break it into bits and bytes in various ways. But eventually, as long as all the bits get to the other end and you can validate that they were received correctly, you really shouldn't care what happens underneath, as long as it's efficient and cost-effective.

Q: When you said that if you went back to 1974 you probably would start to put Digital Object Architecture more into TCP/IP, what exactly would you have done differently there?

Bob: No, I wouldn't have put it into TCP/IP. I would have started out from that construct and then seen how it applies in different network environments, because if you're dealing with a system that is interoperable and it knows how to talk by its protocol interface, it will deal with any other system fully interoperably. The question is, how do you get the bits there? Well, TCP/IP was developed to deal with a fragmented Internet system in the middle for all kinds of applications, and one can just use that directly. Or you could imagine rethinking the underpinnings of the Internet if you wanted to. But remember, the Internet's everywhere around the world and it's unlikely you're going to get everything to change—certainly not all at once, and maybe never for everything. So you need to deal with all of the vagaries, including what you have in today's Internet, going forward.

Q: How different is the Internet today from what you imagined it would be?

Bob: Well, it's very different in the sense that I saw this as a research experiment when we started. We wanted to see how you could get different networks to work together and get the computers on those nets to be able to talk with each other, so we were thinking a small number of networks, a small number of computers. Remember, when we started this, there were no workstations or PCs or any of that. The only things you had that were viable to even work with were the large time-sharing systems. These were all multimillion-dollar machines, and very few people in the world had them—maybe 100 institutions when we started. So that was what we were looking at, and little by little it kind of grew. And then suddenly there were workstations around, and by the early 80s, personal computers started to appear. And then instead of having 100 machines, we were thinking maybe thousands, tens of thousands, hundreds of thousands even, and today what do you have? Probably 3 billion devices on the net and the possibility of 20 or 100 billion with the Internet of Things.

Q: So there's been a huge scaling up of the number of devices connected.

Bob: Well, we've got a lot more things connected than before. Sometimes, even in one machine, you've got lots of virtual machines and many applications running in the same device. I think it's also fair to say you've seen a scale-up in bandwidth. We were starting with 50 kilobits per second, and I think it's not unreasonable to assume that we're dealing in the range of between 10 and 100 gigabits per second today, on average, so that's a factor of about a million in scale-up in terms of memory. And in terms of computational power, we had machines that would probably be outdone by the cheapest of digital wristwatches today, so we've got an increase of at least a factor of a million in computational speed. And it doesn't show any signs of slowing, so maybe 10 years from now it'll be a factor of a billion, and there's no other technical contribution in the history of technology that has scaled over that range of possibilities.

Q: Indeed. In this very impressive process of scaling, what roles do you think mathematics has played in the analysis and design of such systems?

Bob: Well, that's a very interesting question. I started out my career with the intention of an academic career, and I guess I was pretty much trained along traditional mathematical lines. I was really intrigued by problems in mathematics, and I was pretty good at dealing with some of them. There were a lot brighter people who were dealing with really complex theoretical math problems. But I found the applied ones that I was dealing with really intriguing, and I just enjoyed working on them, and there have been results having to do with networks that are interesting results that have come out over the years. Little's Theorem is an example of one. There's the work that [Len] Kleinrock did on the independence assumption, where he was able to prove some things about network performance by assuming certain things, which he couldn't prove were independent, but just by assuming that they were, he could get a closed-form solution.

There was a paper that was written by Len Kleinrock and Howard Frank and me. Howard did all the early topological design for the ARPANET. The three of us wrote a paper, and it was called "Lessons Learned in Computer Communications Theory and Design." One of the questions we asked at that time was: How big a network do we know how to design and build? And we all thought about that, and each of us had our own separate opinions, and it was all about a 60- or 64-node network. We all came and we said, that's amazing. We all have the same view on that, and why is it? Well, Len chimed in first, and he said, well, when you solve the equation for throughput versus delay in networks, the denominator of the equation goes to zero and the number of nodes is about 60, so therefore we can do up to 60, because this equation goes to infinity. And Howard said, well, that's really interesting, because I didn't pay any attention to the math, but I do a lot of simulation work, and by the time you're up to 60 nodes, I run out of memory space in my simulations to actually do the computations anymore. And I said, that's really interesting, because having just built the ARPANET—this was 1972—I said, we're running on 50-kilobit lines today, and if you take a look at the routing tables, if the routing tables got to be much bigger than 60 or 64 nodes, every bit of bandwidth in the network would be consumed passing the routing tables around, and so we just don't know how to do them. Now, of course, it's true if we went to megabit lines, then that constraint would go off the books, and if we had more memory, Howard's simulation wouldn't have failed. Len's equations probably would have had to be tuned in a different fashion. But the bottom line was, here were three different ways of viewing it, and it turns out that the implementation really didn't depend a lot on the simulation, which didn't really depend a lot on the mathematics, so they were just three different ways to attack a problem.

Now, I can't speak for every possible application, and I think we may get more and more cases in the future where you really do need good mathematical models to know if things will even work, but I don't yet know what they might be.

Q: Thanks, Bob, for sharing your insight with us.

PART VI
END TO END

To your computer, the Internet sometimes looks like a black box. It transmits and receives your messages without really knowing what's going on inside the network between you and the others you are communicating with.

End-to-end control is our sixth and final networking principle. In chapter 13, we'll see how Internet devices use feedback provided from receivers to infer and manage congestion in the network. This calls for distributed coordination among devices, sending each other acknowledgment messages back and forth through the network when they properly receive packets.

Congestion control will conclude our three-chapter discussion of the inner workings of the Internet. In chapter 14, we will turn back to social networks, looking at how people can still be connected even at seemingly opposite ends of the network. As we will see, it calls for people to discover short paths using only local information, making it a small world even in such a huge and constantly growing network.

13

Controlling Congestion

When discussing packet routing in chapter 12, we never said anything about how many packets could be in transit at a time. Since the hardware making up the links in the Internet can only send bits so fast, there needs to be a way to control the traffic. The transport layer is tasked with making sure the demand for the Internet doesn't exceed the supply on its links, and when it does, it has to run **congestion control** at the end hosts to bring the demand back down to a tolerable level.

HOW TO CONTROL CONGESTION?

Take a look at illustration 13.1, which shows an example of congestion on a link. Alice is currently sending data at 30 Mbps, while Bob is sending at 20 Mbps (remember that Mbps is a measure of speed, in millions of bits per second). The link capacity is 40 Mbps—meaning it can transmit up to 40 million bits in 1 second—while the users are together demanding 50 Mbps, which is 25% larger. Demand is exceeding supply.

So, what happens? Packets start to accumulate in the buffer at the front side of the link. They are stored there, forming a queue, waiting for a chance to be transmitted. It's like being stuck in a traffic jam (illustration 13.2), with the volume of cars on the road causing everyone to have to inch forward slowly until they reach the front of the line and can drive on. The time that a person sits in a traffic jam is analogous to the time a packet waits in a queue before transmission (though obviously on much different timescales, minutes vs. milliseconds).

To make matters worse, as more cars enter a traffic jam, the line becomes longer for the people who enter it later. Similarly, as more packets pile into a buffer, the queue gets larger, creating even more congestion. Eventually, the buffer will overflow, causing packets to get dropped, like water spilling

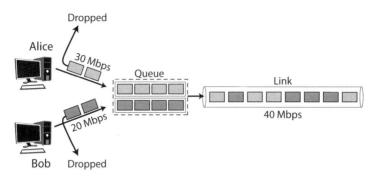

Illustration 13.1 The two users' collective demand for the link, 30 + 20 = 50 Mbps, is higher than the link's capacity, 40 Mbps, causing congestion.

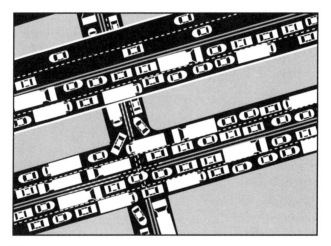

Illustration 13.2 Congestion in the Internet is similar to what you experience in a traffic jam.

out the top of an already full bucket. The specific packets that are dropped depend on the details of the queue-management protocols.

How can the Internet (specifically, the transport layer) control this?

Computer scientist Van Jacobson proposed the first congestion-control mechanism in the late 1980s. It was named **TCP Tahoe**, since it would be part of TCP—the dominant transport layer protocol—and after Lake Tahoe in the Sierra Nevada (as we'll see, many congestion-control algorithms are named in this fashion). It first became part of TCP in 1988 and is thought to have saved the Internet from collapsing in the late 1980s and

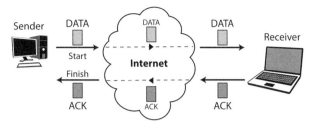

Illustration 13.3 When a sender transmits a packet, the receiver sends an acknowledgment (ACK) on proper receipt. ACKs are negative feedback signals that give senders a measure of network congestion.

early 1990s. TCP Tahoe has been studied extensively and improved several times since then, but many of the essential ideas for congestion control on the Internet were already there in its first versions.

Feedback Once Again

Remember back in chapter 2 when we talked about WiFi devices sending acknowledgments? With TCP, the Internet's **end hosts** follow a similar system: for every packet a transmitter sends over the Internet, the receiver will reply with an acknowledgment (ACK) packet back to this sender when (and if) it gets there successfully. You can see this idea in illustration 13.3: when an end host gets the ACK, it knows the delivery of that packet was successful. Otherwise, after a certain amount of time, it will try to resend the packet.

While promoting reliability, you can imagine that acknowledgments by themselves could actually make congestion worse. In a congested network, packets may be delayed a long time or dropped altogether, causing the receiver to not send an ACK. Without ACKs, senders will resend the packets, adding more to the buffers in this already congested system. When these retransmissions don't get acknowledged either, they will try to send again, and so on, leading to a vicious cycle.

But the ACK system also provides a clever way of managing congestion. Transmitters can use acknowledgments (or the lack thereof) to infer the conditions in the network, since these ACKs indicate if and how long it's taking packets to reach their destinations. If packets are arriving successfully, then they can keep sending, and perhaps at an even faster rate, since rapid acknowledgment may be a sign that the paths are not being used to full capacity. But these acknowledgments getting jammed up or lost is a sign that the network is congested, so once this happens, the transmitters should reduce their sending rates to relieve congestion.

In this way, an acknowledgment can serve as a feedback signal to a sender, indicating when transmission rates should be lowered or raised. We've seen negative feedback pop up as an important theme of networking many times in this book, especially in part I, where it helped cell phones to control their power levels (chapter 1), WiFi devices to back off their transmission rates (chapter 2), and ISPs to regulate data demand (chapter 3). We saw its counterpart, positive feedback, in our discussion of how to trigger information cascades in chapter 9. Having negative feedback in the Internet is important for preventing congestion collapse, too. Negative feedback is often a "positive" thing!

Now, what about a congestion-control mechanism where the Internet's routers, rather than the end hosts, are tasked with deciding what rates the transmitters should use? This may sound intuitive: after all, the routers are the ones actually in the network, so they could probably infer the link-by-link congestion well. But if routers managed congestion, they would be required to keep tabs on the end-to-end connections. This would also violate the **end-to-end design** philosophy of the Internet that we alluded to in chapter 11: leave the jobs that naturally fit the end hosts to the end hosts. With TCP, congestion control is managed by the end hosts.

Sliding the Window

With a feedback mechanism in place, the next step is to decide how end hosts will use it to regulate their transmissions. Perhaps we could have an end host send a packet and then wait for the ACK of that packet before sending another? A "one-for-one" scheme would be pretty slow and inefficient, since it would mean each end host could only have one packet in the network at a time.

TCP pipelines this by providing a bigger allowance than just 1. Each sender maintains a **congestion window** that places a limit on the number of outstanding/unacknowledged packets that it can have in the network at a time. The larger the window size, the more outstanding packets are permissible: if it's 3, then up to three packets can be sent before the sender has to pause and wait for acknowledgment packets to come back. You can see this in illustration 13.4: for each new ACK received by the sender, the

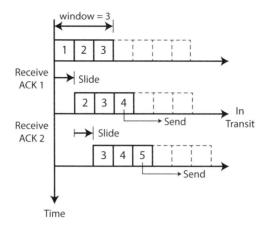

Illustration 13.4 For every properly received acknowledgment, the congestion window slides to the right by one packet, allowing the end host to send another packet into the network.

window slides one packet to the right, allowing the sending of a new packet. Because of this behavior, it is sometimes called a "sliding window."

TCP's sliding window is similar to Netflix's DVD rental policy mentioned at the beginning of chapter 7. Netflix allows renters to have a fixed number of movies out at a given time, depending on how much they pay per month. The more they pay, the more movies they can have out at once, giving them a larger "window." A person at her maximum allowance has to return a DVD before taking another one out.

Growing and Shrinking the Window Size

If the congestion window is the means for regulating traffic, then how can it adapt to network conditions?

If a transmitter doesn't sense any congestion, the window size should be allowed to grow, to make optimal use of the network's supply. Otherwise, we won't be fully taking advantage of the efficiency possible through packet switching. Let's be clear: increasing the window is different from sliding the window—it gets larger in addition to sliding forward. In the Netflix DVD analogy, sliding the window is like returning a DVD; increasing the window size is like subscribing to a higher quota DVD service.

A common goal of TCP is to increase the window linearly—by 1—when the current windowful of packets is properly received. So if the window size

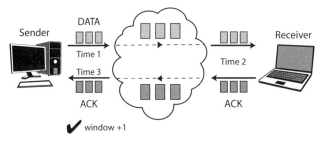

Illustration 13.5 Initially (Time 1), the window is 3, so the sender transmits three packets over the network. The receiver gets each of the packets (at Time 2), and sends out acknowledgments one by one. When the sender receives three acknowledgments (at Time 3), its window increases to 4.

is 3, once the sender receives acknowledgments for three packets, the window may increase to 4. Then, once four more packets are acknowledged, it will increase to 5, and so on, as in illustration 13.5. For more information about how to achieve a linear increase like this, check out Q13.1 on the book's website.

In contrast, when there is congestion, the window size 'should be reduced. TCP usually tries to decrease it multiplicatively—rather than linearly—for the next windowful of packets it sends, typically cutting it in half to be exact. So if an end host detects congestion when its window is 8, it may go down to 4. If congestion is still sensed in the next packets, it may be reduced to 2, and so on. This multiplicative decrease is the same order of magnitude that we saw in chapter 2 for WiFi devices backing off after a collision. Remember that there's a window used to regulate WiFi transmissions too, except a higher contention window for WiFi means less frequent transmissions, whereas a higher congestion window for TCP means a higher transmission rate.

Increasing the congestion window size additively and decreasing it multiplicatively means that the control of packet injection into the network is conservative. It would have been much more aggressive if it were the other way around: multiplicative increase and additive decrease. Actually, before the **congestion avoidance** phase, TCP usually has a **slow start** phase in which the window size is increased more aggressively, when the connection between the end hosts is first established. "Fast start" would probably be a more fitting name for this period of time, because during it, the window is ramped up quickly to a reasonable value.

Now, how exactly do end hosts infer whether there is congestion or not in the first place? After all, they really have no idea what the network looks like. They have no idea what paths their packets are taking, which other end hosts share links with them, or which links along the paths are congested. They only get an idea of what's going on through the acknowledgment packets they receive, and yet they each have to make an informed guess: is their connection experiencing congestion somewhere in the network? This is a challenge of distributed coordination in networks!

HOW TO INFER CONGESTION?

Many congestion-control algorithms have been proposed over the years, and several of them have been implemented in widely deployed systems. All the main ones use a sliding window controlled by negative feedback to regulate transmission rates. Where they begin to diverge is in how congestion is inferred, which has different implications for window size updating.

Packet Loss as the Signal

The earliest versions of congestion control, TCP Tahoe in 1988 and its slightly modified variant **TCP Reno** in 1990 (named after Reno, Nevada, a city near Lake Tahoe), made an important assumption: if packet loss occurs, then there's congestion. This sounds reasonable. At first glance, it also sounds pretty easy to implement: since a successful acknowledgment means a packet was delivered, then the lack of an ACK should mean it was lost. But how can we tell for sure that no ACK has been sent? Maybe it's just delayed, or maybe an ACK was sent and it got lost on the way back.

TCP uses two common-sense approximations to make a reasonable guess as to whether a packet has been lost. The first is: if the sender waits a long time and the acknowledgment for a packet doesn't come back, the packet was probably lost. How long is "long"? TCP keeps a timeout counter based on **round-trip time**, or RTT, between the sender and the receiver. The RTT is the time it takes for the packet to reach the receiver plus the time it takes for the acknowledgment of that packet to get back to the sender. The timeout counter could be something like three times the normal RTT, to be confident that a sufficient amount of time has been allowed to see if the packet was just delayed.

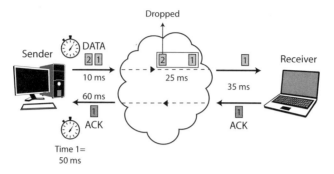

Illustration 13.6 A packet that remains unacknowledged after an unusually long time, say, three times the normal round-trip time, probably has been lost. In this example, Packet 1 is acknowledged 50 ms after it was sent, while Packet 2 will eventually time out and be declared lost.

What is the "normal" RTT? It's the round-trip time for a packet that is experiencing reasonable network conditions (i.e., not much congestion). The smallest of the recent RTT values that the transmitter has experienced can be taken roughly as the no-congestion, normal round-trip time.

You can see an example of this first rule for inferring congestion in illustration 13.6. Ten milliseconds after the sender and receiver establish a TCP connection, the sender transmits two packets. Along the way, at 25 ms, the second packet gets delayed and ultimately dropped. The first packet makes it to the destination after 35 ms though, and the sender gets an acknowledgment for it at 60 ms. The RTT for this packet is the difference between the transmit and acknowledgment times: $60 - 10 = 50$ ms. The sender may use the average of some recently observed RTTs (like this 50 ms) as the normal RTT. After maybe three times that value, say 150 ms, Packet 2 will time out, and the sender will assume (rightfully so) that it was lost.

The second approximation is: if the sender has received acknowledgments for several packets that were transmitted after one it's still waiting for, then this unacknowledged packet was probably lost. Since each packet is assigned a sequence number before transmission, TCP can keep track of the transmission order. Using this, it can tell that Packet 1 was sent first, 2 was second, and so on.

You can see an example of this in illustration 13.7, where the sender is waiting for an acknowledgment for Packet 9. It starts to get acknowledgments back for a few of the packets that it transmitted after 9 (numbered

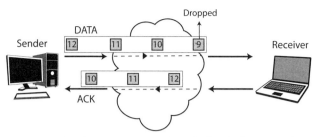

Illustration 13.7 A packet that remains unacknowledged after several subsequent packets have been acknowledged probably has been lost. In this example, the sender gets ACKs for three packets after the ninth one in the sequence, so it may assume that Packet 9 was lost.

10, 11, and 12), but still nothing from the ninth one. Of course, Packet 9 may have traversed a different path with a longer RTT. But if the ACKs for as many as three later packets have already arrived, chances are that the packet is not just late, but lost. These are pretty simple but smart heuristics for distributed coordination.

Inferring congestion based on lost packets was the standard for almost a decade. This method doesn't take into account a couple of factors, though. For one, there's something besides congestion that can also cause packet loss: poor channel quality. This is especially true for wireless networks. Remember in part I when we talked about cell phone signals interfering with one another and WiFi transmissions colliding? Both of these can cause significant packet loss, but they aren't indicative of congestion inside the Internet. There have actually been many proposals for TCP wireless to mitigate these problems.

   Another issue with loss-based inference is that by the time packets are being dropped, it's often too late to react to congestion anyway. Many of the remaining packets on the congested paths may also start to drop, forcing the senders to start over. This can be mitigated by switching to another signal for inference, which we discuss next.

Packet Delay As Feedback

After Tahoe and Reno, **TCP Vegas** was invented in 1995. Besides being named after a different place in Nevada (this time its largest city), TCP Vegas introduced a new paradigm for controlling congestion: using packet delay, rather than packet loss, as the inference signal.

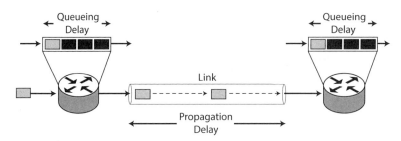

Illustration 13.8 A packet's round-trip time has two major components: propagation delay and queueing delay.

How can delay be measured? Maybe a transmitter can compare the RTTs of packets to the normal RTT discussed before. This seems valid (and indeed it is), but the reason goes deeper, because two (main) types of delays impact a packet's RTT while only one of them is indicative of congestion. You can see these components in illustration 13.8. Even if there were no other packets in the network, there would still be a **propagation delay** associated with getting across the links from the sender to the receiver. How fast a bit of information can get from one end of a link to the other isn't really dependent on congestion; it's instead dictated by the quality of the hardware that the link is built of and is limited by the fundamental laws of physics. What really changes with congestion is the **queueing delay**, or the time a packet spends waiting in router buffers in between the links. The heavier the congestion is, the longer the buffer lines will be, and the more time each packet will have to spend waiting before a link becomes available.

The key is that we can expect any changes in the RTT to be caused by the queueing delay. The propagation delay on a link will stay roughly the same on a small timescale. This makes delay an accurate signal of congestion conditions.

Packet loss is a binary measure. It gives an end host two possibilities: either there is congestion (packets are being dropped) or there isn't (packets are not being dropped). Delay allows an end host to factor in the extent of congestion. If the delay is a little more than expected, then the window size can be dropped a little. If it's a lot more than expected, it can be dropped a lot. Similarly, if there's less delay than expected, the window size can be bumped up accordingly.

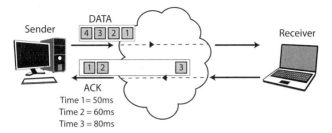

Illustration 13.9 With delay-based inference, the sender would adjust the window size for each packet's measured round-trip time (RTT), whereas for loss-based inference, it wouldn't lower the size until the fourth packet is declared lost.

In other words, rather than waiting until packets are getting lost, delay-based inference allows end hosts to respond to the first signs of congestion. You can see an example in illustration 13.9, where a sender has transmitted four packets to a receiver, in rapid succession to one another. The ACKs show increasing delay in how long these packets take to be received, clocking 50, then 60, and then 80 ms RTTs. The fourth packet has been lost altogether. With delay-based signaling, the sender could respond to this gradually as the RTTs begin increasing, with more drastic decreases in the window size as the RTTs creep up. With loss-based inference, the sender would wait until the fourth packet is lost, which may already be too late to prevent further and more severe delay and loss.

After TCP Vegas, many other congestion-control algorithms have been proposed and implemented in TCP over the past two decades. **FAST TCP** in 2002 modified how the congestion window is adjusted based on estimated delay and helped stabilize congestion control. **CUBIC TCP** in 2005 combined loss- and delay-based signaling and was the default implementation in the Linux operating system until **TCP Proportional Rate Reduction** became the default for Linux in 2012.

Now that we understand congestion control's driving ideas, let's look at examples of two algorithms in the next sections.

TCP RENO: DETAILS OF THE DETECTIVE

A quarter of a century after it first came out, the loss-based congestion-control algorithm TCP Reno is still in widespread use, though some improvements have been made to it over the years. Its main operation is actually quite simple. For each windowful of packets, the sender asks itself: were all of them received properly?

- If they were, then it increases its window size by 1.
- If they weren't, then it cuts its window size in half.

A packet being "received properly" is based on the two approximations we discussed before: the acknowledgment has to come back (i) within a reasonable amount of time and (ii) reasonably in sequence with the original transmissions. By this logic, the sender can assume a packet was lost if either (i) a certain number of RTTs have passed or (ii) if a certain number of subsequent packets have been acknowledged. There are other, more subtle features of TCP Reno, but we won't have time to get into them in this book.

We'll look at an example to see the main operations of this algorithm. For simplicity, let's say that the RTT for each packet is the same, which means that the acknowledgments for all packets in a windowful will be received simultaneously. Time 1 will be when one RTT has passed, Time 2 when another has passed, and so on. In reality, a typical RTT is about 50 ms across the United States, and will obviously vary as the congestion conditions change, as well as from packet to packet.

When the sender establishes its session with the receiver, let's say it starts with a window size of 5:

- At this time, it sends out a windowful of five packets and then pauses.
- At Time 1, the sender gets ACKs for all five packets. It slides the window to the right by five, increases the window size by one to 6, and sends out six packets.
- At Time 2, what happens if the sender gets ACKs for all six? It will slide the window by six, increase the size to 7, and send out seven packets.
- At Time 3, if it's again successful, the window size becomes 8.
- At Time 4, let's say all ACKs come back except for the third one. Since five packets after this one have already been acknowledged, the sender will declare the third one lost. So, it halves the window size to 4, and sends just four packets this time. You can see this in illustration 13.10.
- At Time 5, the sender gets ACKs for these four packets. So it slides and increments the window, and then sends five more packets.

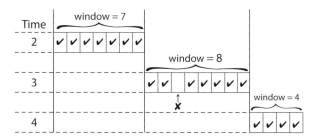

Illustration 13.10 At Time 3, the window size is increased from 7 to 8, and the sender can have eight packets out. When one of these is detected as lost, at Time 4 the window is halved, to 4.

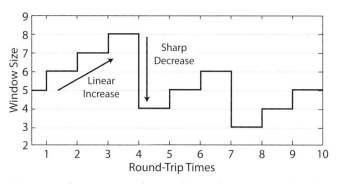

Illustration 13.11 With TCP Reno, the window either increases linearly (no packet loss) or decreases multiplicatively (packet loss).

The evolution of the window size for these first five RTTs, and how it may change beyond that, is shown in illustration 13.11. If you're interested in what could cause the evolution shown here for the remaining four RTTs, check out Q13.2 on the book's website. The bottom line is that when there is no packet loss, the window size grows linearly, whereas when packet loss occurs (at Times 4 and 7), it decreases sharply.

TCP VEGAS: DETAILS OF THE DETECTIVE

For comparison, let's take a closer look at TCP Vegas, the first delay-based congestion-control algorithm. It has been implemented in a few operating systems of computers, including Linux and FreeBSD.

TCP Vegas uses observed round-trip times to determine throughput rates at the current window size, or how many packets are being acknowledged per second. The ideal throughput rate is the one that can be expected if there is (almost) no congestion in the network, giving the minimum round-trip time. For each packet acknowledgment, the end host asks: how does the current throughput compare to the ideal throughput at the current window size?

- If the throughput is too close to the ideal, the end host increases its window size by 1.
- If the throughput is too far, it decreases its window size by 1.
- If the throughput is the desired amount (neither too close nor too far), it doesn't change the window size.

The desired amount is some prescribed threshold, like 3. A difference below 3 is too close, and above 3 is too far.

Eventually—if the algorithm is fine tuned properly—the network should find itself at an equilibrium, where each of the senders are at the desired throughput. That is, until something changes, like a new session being established, at which point this procedure will search for the new equilibrium. In equilibrium, the network is at the "perfect" level of utilization without too much congestion, so the window sizes stop changing. Of course, the reality of network dynamics makes "equilibrium" only an ideal.

The mathematics behind achieving sending-rate equilibria over multiple users is a bit too advanced for this book. In any case, for it to be reached, it is very important that all end hosts are following the protocol. One sender deciding to increase its transmission rate from the agreed-on spot probably wouldn't affect the congestion conditions much, but it would be unfair to everyone else. If the others decided to increase their rates too, the congestion would pick up, and soon they would each be in a worse state than before. This may remind you of when we talked about different methods for enabling efficient sharing of network resources in part I. Controlling transmission powers in a cell and controlling congestion in the Internet serve two different purposes, but their principle of operation is the same: let each device adjust its own level based on feedback it receives about conditions in the network, targeting an equilibrium. Without the proper protocols, the Internet could wind up like the depleted pastures in the tragedy of the commons from chapter 3. Illustration 13.12 shows the distributed

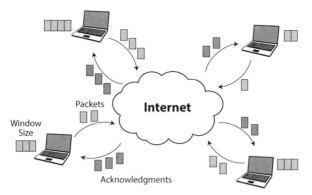

Illustration 13.12 Congestion control is accomplished in a distributed manner, with each end host controlling its window size through negative feedback.

nature of congestion control, which you can compare to illustration 1.19 in chapter 1.

Finally, let's walk through an example of TCP Vegas. For simplicity, let's say that the minimum RTT never changes and is fixed at 50 ms. In reality, this is not the case, and the sender will dynamically adjust this minimum value based on previous measurements. We'll also think of time units as when packet acknowledgments arrive, so each step corresponds to a window-size update.

When the session is established, let's say the sender starts with a window size of 5, as in the Reno example:

- Initially, the end host sends out five packets and then pauses.
- At Time 1, the first packet comes back, and the RTT is found to be 51 ms. The current throughput is

$$5 \text{ packets}/51 \text{ ms} = 98.03 \text{ packets/sec}$$

and the no-congestion throughput is

$$5 \text{ packets}/50 \text{ ms} = 100 \text{ packets/sec}$$

The difference is $100 - 98.03 = 1.97$, which is less than 3, so the window size is increased by 1, to 6. Since there are now four outstanding packets, the end host will send two more, for a total of six. You can see this in illustration 13.13.

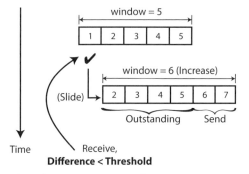

Illustration 13.13 When the congestion window size increases, the end host sends out two more packets, for one extra than before the last packet came back.

- The second packet is back at Time 2, with an RTT of 50.5 ms. What are the throughputs? The current throughput is 6 packets/50.5 ms = 118.81 packets/sec, and the no-congestion is 6 packets/50 ms = 120 packets/sec. The difference of $120 - 118.81 = 1.19$ is again less than 3. The window size is increased to 7, and two more packets are sent.
- At Times 3 and 4, the third and fourth packets come back. Let's say that their RTTs again cause the window size to increase. Then the window will be 9.
- At Time 5, the fifth packet has an RTT of 50.8 ms. This makes 9 packets/50.8 ms ≈ 177 packets/sec as the current throughput, and 9 packets/50 ms = 180 packets/sec as the no-congestion throughput. The difference is roughly 3, so the window size stays the same, and the end host sends out one packet. You can see this sliding window in illustration 13.14.
- At Time 6, the sixth packet comes back, with an RTT of 51.8 ms. What are the throughputs? The current is 9 packets/51.8 ms = 173.7 packets/sec, and the no-congestion is the same as before, 180. This difference, $180 - 173.7 = 6.3$, is now greater than 3. So, the window size decreases to 8, and no packets are sent (see illustration 13.15).

The congestion window for each of these RTTs, and how it may change beyond that, can be seen in illustration 13.16. If you're interested in what could cause this behavior for Times 7–13, check out Q13.3 on the book's website. There's a clear pattern of rise, steady, fall, steady, rise, and so on, as we saw for the transmit power levels in our power control example in chapter 1 (see illustration 1.18). We expect this behavior for delay-based

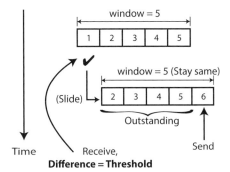

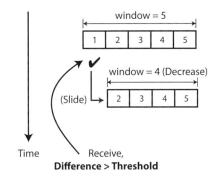

Illustration 13.14 When the congestion window size stays the same, the end host will send out one packet.

Illustration 13.15 When the congestion window size decreases, the end host won't send out any packets.

congestion control: if the network is underutilized, the RTT is too small, indicating the end host could be sending more. If the network is overutilized, the RTT is too large, indicating the end host should send less. When the difference is the same as the threshold, the window size won't change, like how a cell phone under distributed power control won't change its transmission power while it's getting its desired signal-to-interference ratio. Negative feedback is powerful!

To summarize the behavior of these algorithms, illustration 13.17 shows typical evolutions of TCP Reno and TCP Vegas congestion window sizes over time. You can see that TCP Reno, which uses loss-based signaling, is repeating the process of ramping up to a rate that is too aggressive (leading to congestion on the network links) and then cutting back multiplicatively to a much lower level (leading to underutilization of the network

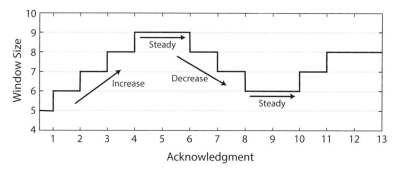

Illustration 13.16 With TCP Vegas, the window size increases by 1, decreases by 1, or stays the same with each received acknowledgment.

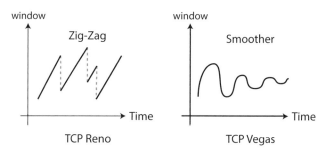

Illustration 13.17 Typical patterns of the congestion window values for TCP Reno and TCP Vegas. The zig-zags between overutilizing and underutilizing the network resources in Reno tend to be smoother and smaller in Vegas.

resources). This "zig-zag" is generally reduced and smoothed out when we use TCP Vegas, a delay-based congestion-control algorithm. While loss is a binary "yes or no" signal, delay tends to allow us to react to congestion more quickly and smoothly, as long as the algorithm's parameters are tuned properly.

Over the past three chapters, we've walked through some remarkable intellectual landscape behind the Internet. We started by looking at the three fundamental concepts behind its design: packet switching, distributed hierarchy, and modularization. Then we went into detail about two of the major functions the Internet has to handle: routing, which is implemented hop-by-hop within the network, and congestion control, which is carried out end-to-end by the devices at the edge of the network. As the "thin waist" of the protocol stack, TCP/IP glues the functional modules below it, like the physical and link layers, to those above it, like the application layer. As part of that thin waist, congestion-control design in TCP and routing design in IP led to a great success. The fact that the Internet has not collapsed, despite its constant growth in geographical spread, number of functions to handle, and demand to manage, is partially attributed to these capabilities.

The goal here was not to master mathematical models or engineering details. For that, there are full courses and even entire degree programs dedicated to the layers in the protocol stack. Rather, we tried to highlight the underlying principles behind the Internet and how routing and congestion control work.

14

Navigating a Small World

The Internet and the social networks it facilitates are huge, and each of us is only directly connected to a tiny sliver of it. As we saw in chapter 10, the average Facebook user was "friends" with about 350 of the 1.65 billion people on the site in 2015. Yet somehow, strangers tend to be connected through surprisingly small short paths, the average degree of separation between two people on Facebook being less than four hops. How is it that such short paths can exist naturally end-to-end between people on opposite sides of the network? As we shall see, this depends on the ways in which social networks are structured and on the ways in which people search for short distances.

IT'S A SMALL WORLD, AFTER ALL

In 1967, American social psychologist Stanley Milgram ran an experiment that kicked off the **small-world** phenomenon, perhaps more widely known as the **six degrees of separation**. It has become one of the most widely told—and sometimes misunderstood—stories in popular science books.

Why Six Degrees?

Milgram asked about 300 people living in Omaha, the largest city in the Midwestern state of Nebraska, to transmit a passport-looking letter to a person living in a suburb of Boston, Massachusetts, on the East Coast. The recipient's name, address, and occupation (stockbroker) were shown on the envelope. Participants were given one important rule: they could only forward the letter to someone they knew on a first-name basis. So if they did not know the recipient by first name (which almost none of them did), they had to send the letter via others, starting with sending it to a friend (one hop), who then sent it to one of her friends (another hop), until the letter

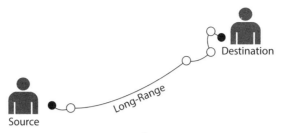

Illustration 14.1 A short path between two faraway people that don't know each other may include one or more long-range links.

finally got to someone who knew the stockbroker by first name. Milgram wanted to know: how many hops would this process take?

You can imagine that knowing the recipient's name and address was important in this experiment. With this information, a person could reason, "well, I don't know the recipient personally, but I know someone who lives near Boston, so I'll forward it to him." It turned out that knowing the recipient's occupation played an important role, too, so that the letter could be passed to people in the same or similar professions. You can guess that the paths these letters took would tend to look something like illustration 14.1: a couple of long-range links to get the letter to the neighborhood of the recipient, and a couple of short-range links based on more specific, local decision making, "distance" being some combination of geographic and occupational spread. Long links from the source followed by short local links to the destination probably remind you of our discussion on Internet routing in chapter 12 and the analogy to the postal mail system, although there are crucial differences in how routing is carried out and evaluated here.

What was the result? Of the letters Milgram supplied, 217 were actually sent out, and 64 of those—about 30%—arrived at the destination. The other letters might have been lost along the way and needed to be considered carefully when analyzing the data. Compared with a later replica of the experiment in the early 2000s via email, which had only a 1.5% arrival rate, the 30% success rate that Milgram got is actually quite impressive. The different number of hops that each letter took is shown in illustration 14.2: the average was only 5.2, and the median (middle of the numbers) was just 6. This is how "six degrees" got its name: a remarkably short distance between people who don't know each other!

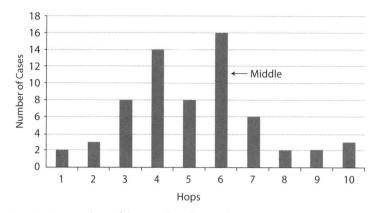

Illustration 14.2 Number of hops taken by each letter reaching the destination in Milgram's experiment. The median (middle) number was 6.

Researchers have long suspected that the **social distance**, or the average number of hops it takes to reach anyone in a population, grows very slowly with the population size. Milgram's results are evidence in favor of that. Over the years, many have raised objections to the conclusions of his experiment, claiming that a different result would be arrived at depending on who was involved, the setup, and other degrees of freedom. But from the 1970s to the online social media era, much empirical evidence has suggested a small world too, from the study of Erdos number among research collaborators to co-star relationships among actors. Other large-scale research studies have also landed on or around the particular number six as the average number of hops, like Duncan Watts's email experiment in 2001 and a study of instant messages between Microsoft Messenger users in 2008.

The degree of separation between Facebook users was found to be just 3.57 in February 2016, down from 4.74 in 2011 and 5.28 back in 2008. Apparently, our already small world is getting even smaller online! (Although not all Facebook friendship would qualify for "know-by-first-name" friendship.)

The small-world concept has permeated popular culture, appearing in movies, TV shows, and songs. A movie titled "Six Degrees of Separation" came out in 1993. In 2007, a website tracking the "Six Degrees of Kevin Bacon" (http://oracleofbacon.org/movielinks.php) was launched based on the idea that anyone in the Hollywood film industry

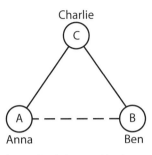

Illustration 14.3 If Anna and Ben both know Charlie, it's more likely that Anna and Ben also know each other. If they do, this relationship becomes a triad closure.

can be linked to the prolific actor through their film roles. In 2015, "Six Degrees of Everything" premiered as a TV show with the premise that the hosts can connect people, things, or both in six steps. These are just three of several examples that you can find on one of the most interesting popular science topics of our time.

Are Small Worlds Surprising?

Should we be surprised by the seemingly universal small-world observation of social networks? On the surface, it seems fascinating that you could reach anyone in six or fewer steps. Thinking about it more, you may reason as follows: if you have 20 friends, and each of them has 20 friends, and so on, then in six steps you could reach $20 \times 20 \times 20 \times 20 \times 20 \times 20$ people. That is already 64 million people. By that logic, six steps should often suffice.

But this logic has a flaw, because it assumes that people's friends don't overlap with one another. You would need each of your friends to have 20 friends that don't include any of your friends, nor any of your other friends' friends. That's obviously not the case: social networks are filled with "triangles," or **triad closures** of relationships. You can see an example of a triad closure in illustration 14.3: if Anna and Ben both know Charlie, it's more likely that Anna and Ben know each other, too.

Six degrees of separation is now truly surprising. But Milgram-type small-world experiments suggest something even stronger: not only do these short paths exist, but they can be discovered by each person using very limited information about the destination and only her own, local view of the network. Compared with sending packets through the Internet in chapter 12, this process of **social search** is perhaps even harder, because people are not programmed to pass messages around to help one another construct a global view. The only useful information they have is embedded in the address and occupation of the recipient, and possibly in the recipient's name, which might reveal something useful about his/her gender and ethnicity. This can at least give the person some idea of which "end" of the network the recipient is on.

From this information, perhaps the participants in these experiments would construct a distance metric between their contacts and the recipient. New York is closer to Boston than Chicago is, on the geographic proximity scale measured in miles, for example. Similarly, a financial advisor is perhaps closer to a stockbroker than a nurse is on some occupation proximity scale, which is a more vague measure but nonetheless can be reasonably quantified. If each person indeed used a simple, "greedy" algorithm based on a local view to forward the letter to his/her friend who is closest to the destination, would that lead to a shortest path?

To summarize, there are two components of the small-world phenomenon:

- The *structural* aspect: Social networks can be formed in such a way that there exist short paths in the first place.
- The *algorithmic* aspect: With very limited local information, a person can find one of these short paths.

Over the years, various models have been constructed to explain both kinds of small worlds. In the rest of this chapter, we build up to one of the most popular ones.

SIX STEPS ARE PLAUSIBLE

How do we build a graph that exhibits a small world? Modeling it will help us understand what networks look like that give rise to the phenomenon in the first place. We need a way to get small path lengths while maintaining a realistic network structure.

Short Distances

We've seen several metrics for describing a network in this book, from simple ones like the number of nodes and links it has to more involved ones like different centrality measures. A common measure used to summarize

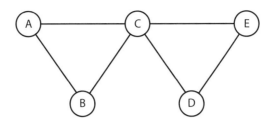

Illustration 14.4 A small example graph used to calculate two metrics: average shortest-path distance and clustering coefficient.

the shortest paths in a graph is its **diameter**, which is the length of the longest shortest path between any pair of nodes in the network. "Longest shortest" probably sounds strange: to determine this length, we find the shortest path lengths between all node pairs, and then take the longest of these.

If the diameter is small with respect to the number of nodes in the network, that would give us what we need for the path-length aspect of small world. But it's actually a bit extreme for our purposes here: when we think about small world, we are more concerned with the average of the shortest paths (remember that six is the middle number in illustration 14.2). If the shortest path lengths for a graph happened to be 1, 1, 2, 2, 2, 3, 4, 5, then the diameter would be 5, which tells us about the worst-case pair of nodes, whereas the average (2.5) or median (2) gives a better summary for small world. You may remember that we used the so-called average shortest path length to compute the closeness centrality measure in chapter 10.

Let's figure out what this length is for the small network in illustration 14.4. The first step is to find all the shortest path lengths. Starting with node A, what is it from A to B? There is a direct link, making the shortest path of length 1. How about A to C? Again, the shortest path is a direct link. For A to D, we have to go over path (A, C, D), which has length 2. How about A to E? It uses path (A, C, E), which also has length 2. You can follow this logic to find the shortest path lengths from B to C (1), B to D (2), and B to E (2); from C to D (1) and C to E (1); and from D to E (1). Out of these ten pairs of nodes, then, the average shortest path length is

$$\frac{1 + 1 + 2 + 2 + 1 + 2 + 2 + 1 + 1 + 1}{10} = \frac{14}{10} = 1.4$$

What does this number tell us? With a judicious choice of path, one person can get to another by taking 1.4 steps on average. It's pretty intuitive in this

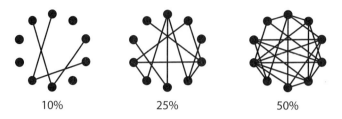

10%   25%   50%

Illustration 14.5 A random graph establishes a link between two nodes according to some fixed chance. As this chance increases, the network is expected to have more links.

example, as the graph has only five nodes, and the shortest paths have lengths of either 1 or 2. As the number of nodes grows, knowing what this length is can be important, and its value can be surprising.

So, what type of network will have a relatively small average shortest path length even with a very large number of nodes?

Random Graphs

Think of a set of nodes that have no links yet. Then someone goes through each pair of nodes one by one, and with some fixed chance between 0 and 100% draws a link between them. Intuitively, the higher this chance is, the more links you will see. You can see some outcomes in illustration 14.5: on the left, the chance of link establishment was only 10%, so only a small number of links were created. As it increases to 50% on the right, many more links are created: roughly half of the pairs are directly connected in this case. The number of links is expected to be proportional to whatever the chance is.

This is how a **random graph** is built. Random graphs keep the shortest path lengths small, because long-range links that connect nodes on opposite ends of the network are possible, and with just a few of them, we can lower the shortest-path distances substantially. However, this process of network formation sounds pretty unrealistic. It would require something along the lines of taking a bunch of people who don't know one another, putting them in a room, and telling them to decide randomly whether or not to become friends. Probably not going to happen, and definitely not a natural way of forming connections.

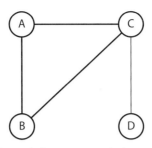

Illustration 14.6 This small graph has one triad closure and five connected triples, making its clustering coefficient 3/5.

Triangles of Friends

In addition to small average distances, a model social network needs to have some tendency for friends of friends to be friends with one another. Random graphs don't provide this, but it's an important characteristic of social networks.

How can this property be measured? By using something called the **clustering coefficient**. It's a measure of how many triad closures we have in the graph, relative to the total number of **connected triples** that may or may not have the last link completing the triangle.

Take a look at illustration 14.6. What are the connected triples? Both (A, B, C) and (B, C, D) are, even though the latter has no direct link from B to D. (B, C, A) and (C, A, B) are too. It may seem redundant to list three connected triples for A, B, and C, but it's not, because each is a distinct path through the nodes differing by one link (for more on this, check out Q14.1 on the book's website). The fact that there are three links between these nodes is what makes ABC a triad closure, while BCD is not.

A graph's triad closures are indicative of the overlaps in who knows whom. The more triad closures the graph has, the more clustering there is. The exact formula for calculating the clustering coefficient is

$$\frac{3 \times \text{Triad closures}}{\text{Connected triples}}$$

Multiplying by 3 accounts for the fact that each triad closure has three connected triples. This makes the clustering coefficient a number between 0 (no clustering) and 1 (complete clustering).

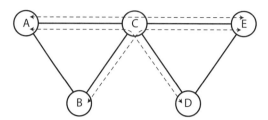

Illustration 14.7 This graph has three connected triples (A, C, D), (A, C, E), (B, C, D), in addition to its two triad closures (indicated by solid lines). This gives a clustering coefficient of 2/3.

What's the clustering coefficient for the graph in illustration 14.6? It has one triangle and five connected triples:

$$\frac{3 \times 1}{5} = \frac{3}{5} = 0.6$$

So, the current connections in this graph exhibit 60% clustering. If there were a link from B to D, the coefficient would jump to 100%.

Let's return to illustration 14.4. How many triad closures are there? Two: ABC and CDE. How many connected triples? Each triangle already has three. Additionally, the graph has three connected triples—(A, C, D), (A, C, E), and (B, C, E)—that are not in triad closures. These are traced out by dashed lines in illustration 14.7, and represent social connections that are not triangles. The clustering coefficient of this graph is

$$\frac{3 \times 2}{9} = \frac{2}{3} = 0.667$$

Random graphs tend to have small clustering coefficients, which is why they cannot explain six degrees of separation. How small is "small"? Without going into detail, it turns out that for a random graph, the formula above roughly translates to the average degree of a node divided by the total number of nodes in the network (remember that the degree of a node is the number of links connected to it). Taking the numbers from Facebook in 2015/2016, this graph would have 1.65 billion users and 350 friends per person on average. If Facebook were a random graph (which it's not), then its clustering coefficient would be about 0.0000002. This is much too small for realistic social networks.

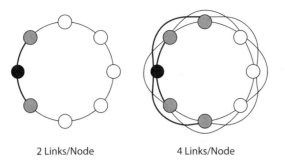

2 Links/Node   4 Links/Node

Illustration 14.8 Regular ring graphs of eight nodes, with different numbers of links per node. As the number of links per node increases, the clustering coefficient increases rapidly.

Regular Ring Graphs

A graph model that can explain small world should have two main properties. One is a small average shortest path length. The other is a large clustering coefficient.

Instead of a random model, what if we consider the other extreme: a very regular structure, like the one in illustration 14.8? This is an example of a **regular ring graph**, in which there is a certain number of nodes (eight in this illustration) on a ring. The structure of the graph is completely determined by two numbers:

- the number of nodes on the ring, and
- the number of links that each node has.

Both will be even numbers, because links spread evenly about each node: half go to one side, and the other half to the other side. When there are two links per node, the graph becomes a pure ring, in which each node is only connected to those next to it. With four links per node, each connects to its two immediate neighbors to the left, and the two to the right. How about for six links per node? There will be three neighbors on each side. The pattern continues in this fashion.

What is the clustering coefficient for this type of graph? For two links per node, it's pretty simple: there are no triangles, so the clustering coefficient is 0. What if we increase the number to four? Now, there's one triad closure centered at each node; at the top left of illustration 14.9, you can see the triangle BCD centered on node C. How about connected triples?

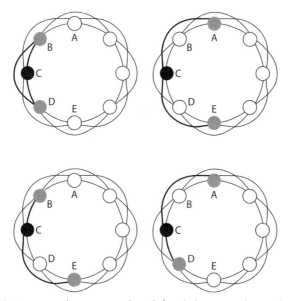

Illustration 14.9 For a regular ring graph with four links per node, each node will have one triad closure (top left) and three additional connected triples (other three pictures) centered around it.

In addition to the three in the triad closure, each node has three triples centered on it; for Node C, (A, C, E), (B, C, E), and (A, C, D) can be seen in the rest of illustration 14.9.

This gives a total of one triad closure and six connected triples so far. Do we need to look at the rest of the nodes? No. Since regular ring graphs are symmetric, the structure around each of the nodes is the same. With eight nodes, for example, there are eight triad closures and $6 \times 8 = 48$ connected triples, which means the clustering coefficient is

$$\frac{3 \times 8}{48} = \frac{1}{2} = 0.5$$

What if there were six nodes instead? $3 \times 6/36 = 1/2$ also. How about 100? $3 \times 100/600 = 1/2$, yet again. The clustering coefficient of a regular ring graph does not depend on the number of nodes, because each additional node adds equal numbers of triad closures and connected triples.

Going from 0 clustering for two links per node to 50% for four links per node is quite a dramatic increase. As more links are added, the growth slows, but nonetheless continues to increase. You can see this trend in illustration 14.10 (if you're interested in the exact equation, check out Q14.2

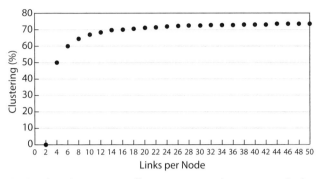

Illustration 14.10 The clustering coefficient in a regular ring graph depends on the number of links per node. As this becomes very large, the coefficient approaches 75%.

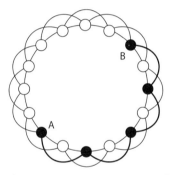

Illustration 14.11 Without long-range connections, regular ring graphs have small average shortest path lengths. In this graph with sixteen nodes and four links per node, it takes four links to go between nodes A and B.

on the book's website): as the number of links per node becomes very large, the clustering coefficient approaches the largest it can be for a regular ring topology—3/4, or 0.75. 75% clustering is nice and large, perhaps even too large.

The regular ring graph model stands in sharp contrast to the random graph model. Unlike the random graph, its clustering coefficient is high, which is realistic for social networks. However, because each node is only connected to its closest neighbors, only short-range connections exist, which makes the regular graph have a large average shortest path length. Several short links have to be traversed to reach nodes on the other end of the network, as you can see in illustration 14.11.

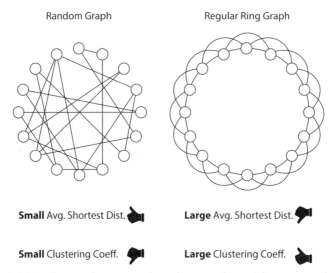

Illustration 14.12 The regular ring and random graph models are in stark contrast to each other regarding the average shortest path length and clustering coefficient.

If we just kept increasing the number of links per node, then wouldn't we establish more direct connections between people, thereby making shorter paths? This would work in theory, but it is also unrealistic, because shrinking the average length more and more would require each person to be linked to larger and larger portions of the population. This is obviously not true in real life, because each person is only "friends" with a tiny fraction of the total population.

The differences between regular ring and random graphs are summarized in illustration 14.12. Is it possible to get a hybrid graph that combines properties of these two types for the best of both worlds: large clustering coefficient and small average shortest distance? If we can do that, we might be able to explain why "six degrees" show up.

Regular Plus Random: Watts-Strogatz Model

Let's think practically here. On one hand, we need a large clustering coefficient. The regular graph provides this quite nicely. On the other hand, we need some long-range links to reduce the distance between the ends of the graph. So, what if we add a few links between nodes on opposite ends?

This is the basic idea behind the **Watts-Strogatz model**, first proposed by Duncan Watts and Steven Strogatz in their 1998 paper published in *Nature*.

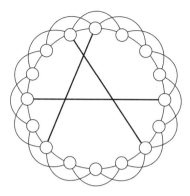

Illustration 14.13 The Watts-Strogatz model adds random links, possibly long-range ones, to a regular ring graph. Whenever a short-range link exists, there is now a chance of establishing a long-range link randomly connecting two nodes.

The model became a very intuitive explanation of small-world networks with large clustering coefficients.

A Watts-Strogatz graph looks something like illustration 14.13. To build it, we start with a regular ring that has a reasonable number of links per node and then randomly add some long-range links between nodes. These random links are generated by going through each link in the regular graph and establishing a link between some random pair of nodes with a certain chance.

The key is that with just a "little bit" of these additional links, we can preserve the large clustering coefficient of a regular ring graph while achieving the small-world effect. With a little randomization, the average shortest path's distance can be reduced substantially. Of course, adding random long-range links will reduce the clustering coefficient, because it will increase the number of connected triples and most likely will not create more triad closures.

How much randomization do we need for distance reduction, and how much can we tolerate while still preserving the clustering coefficient? It turns out that as long as the chance of link establishment is small (e.g., about 10%), the impact on the clustering coefficient will be almost negligible, yet the average shortest-path distance will decrease dramatically. You can see this behavior in illustration 14.14, where we generate the Watts-Strogatz model (with 600 total nodes and six links per node) several times

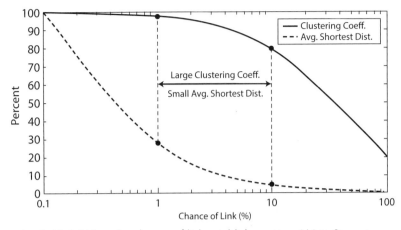

Illustration 14.14 When the chance of link establishment in a Watts-Strogatz graph is small (e.g., 1–10%), the clustering coefficient barely decreases at all, but the average shortest distance drops substantially. This is how the model produces small world with a high clustering coefficient.

for different chances of link establishment, and plot how the clustering coefficient and average shortest path lengths change as a result (divided by their maximum values, to limit the vertical axis to between 0 and 100% for a normed comparison). When the chance is in the range of 1–10%, the distance is small and the coefficient is large, as is needed for a realistic small-world graph.

Why is it that random link establishment reduces the average shortest distance so dramatically but does not impact the clustering coefficient much? The bottom line is the very definition of our metrics. On one hand, the shortest path is an extremal measure: we only care about the shortest of the distances between two nodes, so there's no need to reduce all the paths' lengths. All we need to do is add a few long-range links; even if we add them randomly, the shortest paths will be much shorter. On the other hand, the clustering coefficient is an average measure: it is the total number of triangles divided by the total number of connected triples in the graph. Adding a small proportion of nontriangular, connected triples does not really impact the clustering coefficient.

Therein lies the magic of small world with a large clustering coefficient: we have triad-closure relationships with most of our friends, but a very small fraction of our friends are outside our normal social circle. What Milgram needed to observe six degrees of separation in the first place was this very small fraction of long-range links.

With an understanding of small world's structural aspects, we move now to the second, and even more surprising, part of the phenomenon: how people are able to discover the shortest paths.

MORE IMPORTANTLY: SIX STEPS ARE EVEN LOCALLY DISCOVERABLE

We cannot help but ask: why is it that the shortest path is measured as an extremal quantity, but the clustering coefficient is measured as an average quantity? Rather than finding the shortest paths between all nodes, shouldn't we have to find the average path lengths too? Well, it's generally felt that averaging isn't necessary. For any two nodes in the graph, just the fact that a short path exists should suffice, since the nodes can just use this path to communicate. There's no reason to guarantee that all paths are short, if the others aren't going to be used anyway.

But we have to make sure that people can actually find such a path. This process of social search is not always easy: people don't know the network structure, so if they aren't directly linked to their destination, it can be difficult to determine where to go next. You can see an example in illustration 14.15: if A and C are to communicate from opposite ends of the network, how would they know that they have a common neighbor B they can go through?

As we saw in chapter 12 for the Internet, devices called routers explicitly pass messages to each other based on their own local views, so that they can discover where to send information to place it on the shortest path. This type of routing doesn't happen in social networks. So the question remains: how are people in Milgram-type experiments able to find the shortest routes?

Greedy Social Search

If you were one of those people participating in Milgram's experiment, or one of those along the paths initiated by these people, how would you decide on the best next hop to use just by looking at the destination's name, address, and occupation? You would probably take into account some combination of geographic proximity (relatively easy to see) and occupational

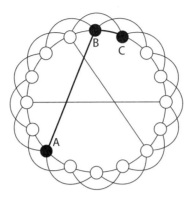

Illustration 14.15 In this Watts-Strogatz graph, a few long-range links are present. A short path exists from A to C: (A, B, C). But how does A know what this short path is, and even whether it exists?

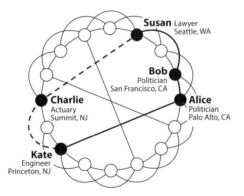

Illustration 14.16 Without a global view of the network, each person may run a greedy social search to determine a path to the destination. This strategy will not always result in the optimal route, but the hope is that the discovered path will have a length that is at least close to the shortest possible length.

proximity (relatively hard to determine) to come up with a "social distance." Then you would look at all your friends whom you know by first name and pick the one closest in characteristic to the destination based on this distance.

This is the idea of **greedy social search**, in which a person makes the best possible decision for sending information from her local information. So if node A wants to get to node Z, she would look at all her neighbors and determine who is the best to send to. Then this node—say, B—would look at the characteristics of Z and determine whom to send to. Then this node (e.g., C) would do the same, and so on, until finally the message arrived at Z. Of course, this approach is not always going to result in the optimal route, since nobody has a global view of the network, but it will often suffice. You can see an example of it in illustration 14.16, where Kate takes path (K, A, B, S) to get to Susan, but the shortest path is actually (K, C, S). Kate runs greedy search to determine that Alice, a politician in Palo Alto, California, is "closer" to a lawyer in Seattle, Washington, than any of her other neighbors. Alice then forwards it to Bob, a politician in San Francisco, who knows Susan directly. In reality, it just so happens that her friend Charlie, an actuary in Summit, New Jersey, has a direct connection to Susan.

Will the average length discovered by this greedy search process be close to the average shortest path length? We hope that it will be, as happens to be the case in illustration 14.16 (2 is not much smaller than 3). If so, then in addition to being plausible, we can say that short paths are discoverable. In the past 15 years, several models for social search beyond the original Watts-Strogatz model have answered this question in the affirmative. If you're interested in these, check out Q14.3 on the book's website.

All in all, it turns out that it is plausible that our world is connected in six steps or less, and that these steps from one end of the network to another can be discovered by distributed, human routing.

Summary of Part VI

End-host intelligence is an important principle appearing in various networking applications. The Internet has often followed the end-to-end design, putting the end hosts in charge of establishing, maintaining, and controlling sessions, while the network itself is only responsible for transferring packets hop-by-hop. In this final part of the book, we first looked at how the transport layer of each transmitting device controls congestion on the Internet, using feedback messages provided by the receivers to infer network conditions and regulate demand for link capacity. Then we looked at how it's still a small world in such a large and constantly growing network, thanks to people being able to discover short paths to others on opposite ends of the network using only local information.

With that, this book's journey through the six principles of networking—sharing is hard, ranking is hard, crowds are wise, crowds are not so wise, divide and conquer, and end to end—comes to an end. In these six parts we've seen so many different networks, talking about how they can be represented both as graphs (e.g., of webpages, devices, or routers) and as functions on top of these graphs (e.g., PageRank, distributed power control, routing).

While this book concludes, we hope that your intellectual journey will continue. You'll continue to come across networks in your life that we did not discuss between these two covers, many of which probably have not been invented yet. When you do, we hope you'll think about which of these six principles might apply to that situation.

Just as important, we hope you'll keep in mind the recurring themes that have come up across these principles. You can think about which of them come up in your daily life, too. Here's a list and brief recap of those six, picked out from the past 14 chapters:

- *Negative feedback* occurs when the network uses signals about its current conditions to drive its state to an equilibrium. This theme came up many times, from cellular power control in chapter 1, to data pricing in chapter 3, to Internet congestion control in chapter 13.

- *Positive feedback* is the opposite, where the network amplifies its effect by feeding off of it, generally moving away from an equilibrium.
- *Distributed coordination* spreads responsibility for a task across smaller entities in the network. Global cooperation can emerge, even though each entity has only a local view of the network. This theme came up many times too, from WiFi random access in chapter 2 to Internet forwarding in chapter 12.
- A *positive network effect* is present when more people joining a network tends to bring benefit to everyone, as in the examples mentioned in chapter 9 and the ratings of products in chapter 6.
- A *negative network effect* is the opposite, where more people join-ing a network tends to detract from everyone's well-being, like the "tragedy" of flat rate from chapter 3.
- *Opinion aggregation* can work remarkably well if people's opinions are unbiased and independent of one another, as we saw in chapter 6. This is why we tend to trust average product and movie ratings more when the average is based on more opinions.

If these themes have stuck with you, then you've learned a whole lot about the power of networks by reading this book!

A Conversation with Vinton Cerf

Vinton Cerf is recognized as one of the "fathers of the Internet." He co-invented TCP/IP with Bob Kahn.

Q: Vint, what was the single most critical decision that you had to make together with your colleagues around 1974 in architecting TCP/IP protocols?

Vint: I think probably the most important decision was to introduce a global addressing structure, which we call Internet Protocol and Internet addresses. And at the time, we already had anticipated multiple networks being part of the network of networks that the Internet forms, and we knew that most of the networks, and maybe all of them, did not have a concept of there being any other network anywhere. Each network thought it was the only network in the world. So, there was no way to say on a particular network, "Please send it to a different network." Bob Kahn and I, when we were doing the design, immediately realized we needed a way to say, "Send this to a network that isn't you in order to allow this network of networks to be formed."

The second thing was not technical. It was a policy decision and that was whether or not we should publicly release the design at whatever level of detail it was at the time and without any restrictions, or patents, or intellectual property constraints of any kind.

We actually thought about that, and recall that we were doing this work for the Defense Department.

The Defense Department had the problem that the only way you could conveniently do networking was to link together similar brand machines. IBM had Systems Network Architecture, which linked IBM machines together; Digital Equipment Corporation had something called DECnet, which would link the Digital Equipment Corporation machines to each other; Hewlett-Packard had something called DS, which I think stood for Distributed Systems and, again, that only connected Hewlett-Packard machines. We thought that the Defense Department should not be put into a position where it had to buy a particular brand of machine in order to achieve a network effect. So that's the first point. That meant a nonproprietary design was needed.

The predecessor to the Internet was called the ARPANET, which stood for Advanced Research Projects Agency Network, and that network was built using dedicated telephone circuits connecting packet switches to each other. It was a uniform homogeneous network with very heterogeneous computers attached to it at various institutions around the United States. This was a homogeneous network of dedicated telephone circuits and packet switches.

When we realized that we were going to have to deal with all of these other platforms, we started developing a mobile radio network and a satellite network in order to cope with mobile vehicles and ships at sea and aircraft scenarios. But then having done that, we realized that we had these diverse packet-switched networks which had different characteristics and we had to find a way to link them all together, so that the computers on any particular network would not need to know how many other networks were part of the system, or how the traffic was being routed. We just wanted it to be like a consumer dropping a postcard in a postbox and expecting that the Postal Service would take care of figuring out where it should go, and how it should get there, and when it should be delivered. So we kind of modeled this after the Postal Service in the sense that the Postal Service works on a global basis. There is a format [for] addressing that is used by the Postal Service to carry messages from one country to another.

Those are probably the most important decisions that were made: release it publicly and remove any barriers there would be to adoption. We decided not to impose any intellectual property constraints on the use of the design. We published all of the documents freely. The institutions that were created later, as the Internet evolved, were similarly very open about their publications and about their specifications, and that is as true today as it was 40 years ago.

Q: A very interesting story, Vint, and very important decisions that influenced how we live and work to this day. It's also interesting that you mentioned those addresses, and I want to follow up. One of these systems is this IPv6, a bigger address space. What is the status of converting to IPv6?

Vint: That's a good question. Let me go back a little bit and say that we went through several iterations of the design of the Internet, four in particular. And the original design had one protocol, Transmission Control Protocol, which managed the flow of traffic from the source host to the destination host across multiple networks. We realized about halfway through the iterations on the design and implementation and test that real-time communication might prove very important. Voice, video, and radar didn't necessarily require a 100 percent guarantee of delivery of the data, especially considering that the data would be refreshed dynamically anyway, especially in the case of radar tracking.

What you want is low delay, so you get an interactive low-latency exchange. If somebody doesn't hear something because a packet gets lost, they're just asked to repeat it. So you hear a popping noise and you didn't hear what that person said, so you say, "What was that? Please repeat."

The same thing could be said for video. You might get a glitch, a part of a frame doesn't show up. But another one is coming anyway, and so don't worry about retransmitting the bits of a previous frame, because the user doesn't want to look at them. If you try to do that (re-transmit), you increase the latency between the parties, and if they're trying to have a video conference call, eventually it becomes untenable because the delays build up.

So instead of allowing the delays to build up, we said, let's allow for packet loss but be attentive to latency. The consequence of that discussion was to split the Internet Protocol [IP] off from the Transmission Control Protocol [TCP] and make it a distinct layer in the system that did not impose the requirement for sequenced reliable delivery. So TCP does the sequenced reliable delivery of data, and the IP just gets datagrams to the destination possibly unreliably. That decision was made around the third iteration of the TCP design.

Now the second thing is we asked ourselves fairly early on, "How many termination points should we anticipate for this Internet?" We really didn't know the answer to that. So the first question we had was, "Well, let's see, how many networks will there be per country?" We thought, well, maybe there will be two national-scale networks in each country, so there'll be some competition. Then we said, "How many countries are there?" And we didn't know the answer, and there was no Google to ask, so we guessed at 128 because that's the power of two and that's what programmers think in.

So we said, "Okay, we need eight bits of information to identify the 256 networks around the world." Then we asked how many computers will be attached to each network. And again we didn't know the answer to that. But we said, well, we should be generous. And so we guessed at 16 million per network, which was outrageously huge at the time, considering computers cost millions of dollars and operated in very big, huge, air-conditioned rooms.

So you could express that number of machines in 24 bits of information. We ended up with a 32-bit address space for IPv4. The aggregate number of termination points that that would give on the Internet was 4.3 billion, and I have to say in 1973 that seemed like an awfully big number for the number of computers that might be on the Internet.

So we proceeded down that path. But then the same year that Bob and I were working on this in 1973, Xerox PARC invented the Ethernet, which was a small piece of coaxial cable with some devices that would allow laptop or desktop computers of the day to be connected together on a broadcast network at very high speeds, three megabits a second at the time, which seemed like a lot. The Ethernet technology proliferated rapidly after about 1980 or 1981 as a commercial offering. Those networks rapidly filled up a lot of the address space on the Internet. The consequence was, around the early 1990s, that it became apparent the initial address space was not going to be sufficient to serve the requirements.

There was a big debate about how much more address space might be needed. We thought the 32-bit address space was inadequate, so should we increase the size to 128 bits, or should we go to a variable length addressing, which is what your phone numbers tend to be. There were reasons not to go down the variable length path. So we ended up with 128 bits, and we called that version of the Internet Protocol IPv6.

At this stage of the game here in 2015, as we're having this conversation, the version 4 [IPv4] address space has been exhausted largely on the net. There is no more address space available. That's a slight overstatement, because some of the parties responsible for assigning Internet address space, the so-called regional Internet registries, still have a certain amount of addresses left to them but not much. Most of them have already run out. There is no new version 4 address space available.

So after the early '90s, realizing that we were going to get in trouble, IPv6 was standardized around 1996. And we thought wrongly that everyone would recognize how important it was to quickly implement IPv6, so that we wouldn't be constrained anymore by the limitations of IPv4, and that didn't happen. People had not run out of address space and only recently ran out, around 2011. And so the pace of implementation of IPv6 capability in parallel with v4 was very slow. Now we're starting to see it pick up, because the version 4 addresses are essentially consumed.

So there's an increased pace of IPv6 implementation in parallel with IPv4. We have to run both protocols at the same time in order to cater to the older community running v4 and the newer community running v6. We have to run both protocols concurrently, because they don't interwork with each other, because you can't fit 128 bits of address into a 32-bit space.

Q: Thank you for walking us through the iterations. It shows a lot of very interesting architectural decisions like this split of the layers. You mentioned TCP/IP driven by this latency and error resilience trade-off, and the fascinating story of how scalable the Internet has become over the past number of decades. These days there is a rise of cyber physical systems and smart cities. Some of these physical control systems need to have very small and almost deterministic latency. Do you think that the rise of these cyber physical systems on the edge of the network, or sometimes within the human body, are also going to force us to revisit the latency issue?

Vint: That's an interesting question. I think the answer is maybe, because there are some circumstances where the latency is very well controlled, because the components of the network that you're dealing with are close together. The thing that we have to be thoughtful about is that speed-of-light delay is not going away. So no matter what you do with protocols, there may be an irreducible minimum delay imposed by distance.

The way you overcome the problem of course is to design and build systems that use networking that's nearby as opposed to distant. When you have things like Bluetooth or even WiFi, it's typically the case that the interaction is local and therefore with very low delay. So there you don't suffer a problem.

The other side of that coin is that some applications can be designed to be insensitive to latency, in which case they can work pretty much anywhere. I have seen both of those angles … being addressed in the application space where some applications … are sufficiently insensitive to latency; Email would be a good example of that. But as you say, for certain cyber physical systems (or otherwise sometimes called the Internet of Things), you need to have the low latency to make the application work.

Q: We also mentioned earlier in this conversation that sometimes the "cloud" needs to come down to the ground and become "fog" in order to ensure a very small latency for certain applications. Nowadays, we watch on TV commercials different companies talking about how the Internet of Things is going to revolutionize the way we live and work and how industrial plants and agriculture will work. And you see objects ranging from consumer wearables to the smart cities blueprints. Which of these do you think is more like science fiction and which do you think have legs and actually will be happening?

Vint: Things are happening on all fronts, which is I think pretty amazing. There are a number of devices that people can wear. Google Glass was one of the early examples of that. But Fitbit and other kinds of products that are helping track our bodies' vital signs in response to exercise and so on are very common. That's already happening, no question. There are other devices that are becoming part of our environment, things that allow your laptop to send information to a big television screen, so that you can enjoy a better visual experience; network-enabled thermostats and things like that that will allow you to remotely control temperature or investigate or determine what the state of the house is; maybe the ability to look at video cameras in order to ensure that the house is still intact.

All kinds of things like that are happening, and I think that we're going to see progress literally on all fronts at some pace over the course of the next several decades. So I don't see much of this as science fiction, to be honest. Even things like self-driving cars that we work on at Google are not science fiction anymore, because they're out on the roads actually trying to navigate on their own.

The one thing I do worry about though is that a lot of these devices may be designed and built with relatively small processing power. The question is whether or not there is enough processing power, for example, to encrypt traffic to ensure privacy of data. And we should be very concerned about the safety and security and privacy of these systems, since a lot of the data that's collected could be interpreted and abused. For example, if you just have a bunch of temperature sensors around the house, it's conceivable with continuous monitoring that you can figure out how many people live in the house and where they are and what their normal daily routine is, when they are at home, when they are not at home.

So we have a whole lot of safety, security, and privacy challenges in the cyber physical systems, Internet of Everything space, which is going to require some very serious research and development in order to overcome or to protect against the potential abuses of these cyber physical devices.

Q: The information privacy you mentioned is indeed a rising concern among many people, especially when you may also have physical objects involved. Should we be concerned about that when we go ahead and purchase that connected "thing," not knowing when they'll be out of date on their security features?

Vint: Absolutely, we should be worried about that, and in fact I would argue that we should not be buying, let alone selling, devices that cannot be updated over time in order to repair the mistakes that have been made in the implementation that lead to vulnerabilities. The headline I worry about is things like "100,000 refrigerators attack Bank of America," because they were infected with malware and launched this denial of service attack. We have very serious technical work to do to [ensure] that devices are controlled only by authorized parties, that the data they accumulate [are] accessible only to an authorized party. We will almost certainly end up having to use cryptographic technology in order to do strong authentication and to protect the confidentiality of data that may be moving from a device in a cyber physical system to some other monitoring system, for example.

Q: I do want to ask you more question: There is a huge application ecosystem around the Internet. What is the app that you like most?

Vint: That's a very interesting question, because whenever someone says app, of course I assume that means a mobile. I realize your question is not intended to be quite that narrow.

So the service that I use more than anything honestly turns out to be Google and it's not because I work at Google. It's just because finding stuff on the net is so valuable and so hard, and Google makes it easy and frequently produces the answers that I'm looking for.

I use apps to keep track of the stock market and my own portfolio on a regular basis. I find myself making use of streaming video, either to look at YouTube segments that have been brought to my attention or watch a Netflix movie. Those applications are very popular where you have adequate bandwidth to support the streaming of audio and video.

So for me, these are applications of convenience for gathering information. I read the newspaper online rather than getting hardcopies in the house. And so that's a daily occurrence. I'm a heavy user of email, as are many of my colleagues and contemporaries. Even though there are other communicating alternatives, like texting or Facebook or Google+, I find myself not using those as much as email. And finally, I'm finding in the recent couple of years anyway an increasingly heavy use of video conferencing, which is more or less what we're doing right now. And that is an amazing evolution, because the Google Hangouts, which allow multiple parties to interact with each other simultaneously, almost make it possible to avoid traveling by having meetings with people remotely.

The other thing which I find myself using is the documents of the Google Docs system, the spreadsheets and text documents and presentations. The text documents in particular, in connection with the video conferencing application, have turned out to be an incredibly powerful combination. I find myself working with two or three other authors on papers, where we are simultaneously talking to each other, possibly seeing each other and also editing a shared document concurrently. Rather than having one person in charge of editing the text, anybody is free to make edits and to draw people's attention to them. Because it's a real-time discussion, we're finding closure on the final documents is reached much faster than it would be if we were sending email attachments around. So this ability to work concurrently in a collaborative way in real time is stunningly powerful.

Q: Thank you, Vint, for sharing your insight with us.

Index

0G, 6
1G, 9
2G, 10
3G, 24
4G, 6, 25

access point, 30, 34, 37
ALOHA, 35, 43
amplified, 18
analog, 6
application layer, 223, 225, 227
ARPANET, 213
ascending price auction, 76, 81
attenuate, 7
auction matching, 77
auctions, 75–76, 99, 117
autonomous system, 221

backhaul, 31
base stations, 8, 30
baseline predictor, 137, 145
basic service set, 30
Bayesian ranking, 123, 165
Bellman-Ford algorithm, 236
best-effort, 217
betweenness centrality, 198
binary exponential backoff, 42, 61,
    264
bipartite graph, 159
bits, 9, 230
Border Gateway Protocol (BGP),
    232
bps, 29
byte, 46

capacity, 7, 26, 48, 259
capture, 31
carrier sensing, 33, 38, 243
carrier sensing multiple access
    (CSMA), 33
cells, 7, 30
cellular, 3, 26, 44, 297
centrality, 193, 200, 283
centralized, 23, 33, 243
channel gains, 18
channel quality, 14
circuit-switched, 211, 216, 227, 300
click-through rate, 75
closeness centrality, 194
cluster, 203, 285
clustering coefficient, 285
co-participation, 155, 172
co-visitation count, 172
code division multiple access
    (CDMA), 12, 23, 65
collaborative filtering, 139, 163, 172
collision, 31, 61, 264
community detection, 162
congestion avoidance, 264
congestion control, 223, 259
congestion window, 262
connected components, 98
connected triples, 285
connection-oriented, 224
connectionless, 224
contagion, 201
contention window, 40, 264
converge, 17, 22

cosine similarity, 140
CUBIC TCP, 269
customer-provider, 221

dangling nodes, 98
data frame, 31, 226
data message, 35, 212, 225, 242
data packet, 212, 226, 232, 248, 259,
    265
decapsulate, 226
degree centrality, 193
demand, 7, 55, 66, 79, 217, 259, 297
density, 204
descending price auction, 76, 81
diameter, 283
digital, 9
diminishing marginal returns, 55
directed link, 90, 154, 193, 235
discussion forum, 149, 151
discussion thread, 151
distance learning, 147
distance-vector, 242
distributed, 23, 33, 120, 236, 242,
    272
distributed power control (DPC),
    17, 80, 177, 236, 275
double capture, 31
Dynamic Host Configuration
    Protocol (DHCP), 230
dynamic programming, 236

Emperor's New Clothes effect, 185
encapsulated, 225
end hosts, 226, 261, 269
end-to-end design, 257, 262
enterprise social network, 153
equilibrium, 20, 177, 204, 272
Ethernet, 31, 250
extended service set, 30

fallacy of crowds, 174, 183
FAST TCP, 269

feasible, 17
first-price auction, 77
flat-rate, 44, 57
flipped classroom, 153
flipping threshold, 200
fog, 67, 105, 221, 303
forwarding, 232, 242
forwarding table, 233
frequency channel, 5
frequency division multiple access
    (FDMA), 5
full text search, 87

generalized second-price, 83, 99
graph, 90, 99, 142, 155, 190, 234,
    283
graph partition, 193
greedy social search, 294

header, 225
hertz, 5
hidden node problem, 43
hop, 222, 234, 242, 280
host identifier, 230
HTTP, 223
hyperlinks, 90, 172, 234

IEEE 802.11, 28
importance score, 88, 99, 199, 234
in-degree, 91
incoming links, 91
individualizing learning, 150
information cascade, 175
information retrieval, 158
information spread, 173
interference, 14, 31
internet service provider (ISP),
    47, 216, 262
Internet of Things, 66, 105, 216,
    221, 230, 254
intrinsic value, 174
IP address, 226, 229, 250

IPv4, 229, 250, 302
IPv6, 229, 250, 302
iterative, 17, 236

layered protocol stack, 222, 276
link, 5, 90
link layer, 222
link-state, 242
Long Term Evolution (LTE), 25, 65

massive open online course
    (MOOC), 149, 161, 165, 191
Mbps, 29, 215, 223
medium access control, 33
metric-based, 232
mobile penetration, 3
mobile stations, 8
multiple access, 5. *See also* code
    division multiple access (CDMA)

natural language processing, 157
near-far problem, 14
negative externality, 19, 48, 80
negative feedback, 20, 48, 80, 144,
    177, 261, 273
negative network effect, 60
negatively correlated, 140
neighborhood model, 139, 146
neighborhood predictor, 144
net utility, 55, 78
Netflix Prize, 133
network address translation (NAT),
    230
network effect, 174
network layer, 222
network projection, 161
nodes, 90
NSFNET, 214

one-size-fits-all, 150
open auction, 76

Open Shortest Path First (OSPF),
    232, 242
optimal, 17, 37, 145
orthogonal codes, 12
orthogonal frequency division
    multiplexing (ODFM), 25
out-degree, 92
outgoing links, 93
overhead, 226

packet switching, 211, 216, 227
PageRank, 88, 99, 162, 172, 199,
    217
path, 194, 211, 216, 232, 282, 287,
    293
pay-per-click, 72
pay-per-thousand-impressions, 71
payload, 225
payoff, 78
peering, 221
physical layer, 222, 277
policy-based, 232
positive feedback, 173, 177, 204,
    262
positive network effect, 174
positively correlated, 139
prefix, 230
propagation delay, 268
protocol, 213
public switched telephone network
    (PSTN), 3

Question and Answer (Q&A) sites,
    153
queueing delay, 268

random access, 33, 48, 61, 222
random graph, 284
receivers, 4
recommendation system, 114, 130,
    162, 172
regular ring graph, 287

relevance score, 88, 98
resource pooling, 217
root mean square error (RMSE),
    133
round-trip time (RTT), 265
router, 90, 226, 234, 250, 268, 293
router information protocol (RIP),
    232, 242
routing, 222, 228, 241, 249, 271,
    279, 295

sealed-envelope, 77, 117
search ads, 72
second-price auction, 77, 81
segment, 225
sequential decision making, 176
service set identifier (SSID), 30
session, 212, 223, 270
shortest path, 194, 233, 282
shortest-path problem, 235
signal-to-interference ratio (SIR),
    17, 32, 275
six degrees of separation, 278
slow start, 264
small-world, 278
smart data pricing (SDP), 52, 66,
    111
social distance, 280
social learning, 150
social learning network (SLN), 152,
    191, 192, 217
social search, 281
sparse, 91, 131
split billing, 52
sponsored content, 52
statistical multiplexing, 217
streaming, 129
subnet, 230
switch, 226
symmetric, 142

TCP Proportional Rate Reduction,
    269
TCP Reno, 265, 270
TCP Tahoe, 260
TCP Vegas, 267, 273
TCP/IP, 213, 223, 247, 276, 311
throughput, 35, 43, 216, 272
tier-1 ISPs, 220
tier-2 ISPs, 221
tier-3 ISPs, 221
time division multiple access
    (TDMA), 10, 32
toll-free, 52
tragedy of the commons, 58, 272
transmission power control (TPC),
    14
transmitters, 4
transport layer, 223, 259
triad closures, 281
truthful bidding, 79

unicast, 212
usage-based, 47, 66, 79, 177
utility, 54

valuation, 75
Vickrey-Clarke-Groves, 85
viralization, 169

watt, 15
Watts-Strogatz model, 290
webgraph, 90, 154, 234
WiFi, 26, 48, 93, 213, 261, 303
WiFi station, 34
wireless, 3
wireline, 3
wisdom of crowds, 101, 118, 48, 93,
    213, 261, 303

zero-rating, 52